Great Lakes Fish: Illustrated
A Field Guide

John C. Tomikel

Allegheny Press

ISBN 978 – 1975843724
197584372X

Tribute

Professor Paul Thomas (2002 photo)

Among his achievements, Paul Thomas spent years documenting the species of fish in Lake Erie and many of its feeder streams. His research, which was published in 1993 permits other scientists to compare his findings with their own research and make inferences and conclusions about the status of fish and the lake.

When Professor Thomas began his study, the Blue Pike and the Cisco, which once thrived in the lake, had recently been declared extinct. The goby and the paddlefish were not yet in the lake. These are points of scientific interest.

Paul Thomas holds three degrees from Michigan University. His PhD was granted in 1964. He has been a visiting scholar at Harvard University, Cambridge, Massachusetts and a research fellow at California Technical Institute. He joined the department of biology at Edinboro University of Pennsylvania in 1968 and once served as its department chairman. He is now a retired 88 year old.

Contents

I. Facts and Opinions   page 4

II. Fish Families   page 13

    1. Lampreys   2. Sturgeons   3. Gars   4. Bowfins
5. Herrings   6. Trout   7. Smelt   8. Mudminnows
9. Pikes   10. Minnows   11. Suckers   12. Catfish
13. Trout-perch   14. Codfish   15. Killifish
16. Silversides   17. Sticklebacks   18. Temperate bass
19. Sunfish   20. Perches   21. Drum   22. Sculpins

III. Practical Classification   77

1.	Sport Fish	77
2.	Pan and Food Fish	83
3.	Little Fish	107
4.	Lamprey	116
5.	Odd Fish	119

IV. Research Data   134

1. Toledo and Ohio List   134
2. Lake Superior List   135
3. Lake Erie List   119

V. General Index   148

I. Facts and Opinions

By way of introduction.

The **Great Lakes of North America**, are a series of interconnected freshwater lakes located primarily in the upper mid-east region of North America on the Canada-United States border, which connect to the Atlantic Ocean through the Saint Lawrence River. Consisting of Lakes Superior, Michigan, Huron, Erie, and Ontario, they form the largest group of freshwater lakes on Earth by total area, and second largest by total volume containing 21% of the world's surface fresh water by volume. Lake Baikal in Siberia contains 23 % of the earth's fresh water. The total surface of the Great Lakes is 94,250 square miles and the total volume (measured at the low water datum) is 5,439 cubic miles.

The Great Lakes may be referred to as inland seas due to their characteristics of rolling waves, sustained winds, strong currents, great depths, and distant horizons. There have been many shipwrecks due to the rolling waves caused by severe weather and many remnants of these ships are being found each year. Lake Superior is the second largest lake in the world by surface area, The northern half of the Great Lakes is bordered by a sparsely populated Canada while the southern half has a large population often referred to as the Great Lakes Megalopolis.

The Great Lakes began to form at the end of the last glacial period around 14,000 years ago, as retreating ice sheets exposed the basins they had carved into the land which then filled with meltwater. The lakes have been a major highway for transportation, migration and trade, and they are home to a large number of aquatic species. Many invasive species have been introduced due to trade, and some threaten the region's biodiversity.

Great Lakes fisheries are under stress and new approaches must be found to deal with these challenges. The list of needed research encompasses fish biology and population dynamics, habitat and ecosystem health, toxic chemical contaminents, the potential effects of climate change, socioeconomic impacts and conflict resolution between and among the many entities concerned with the welfare of the lakes and legally able to do something about it.

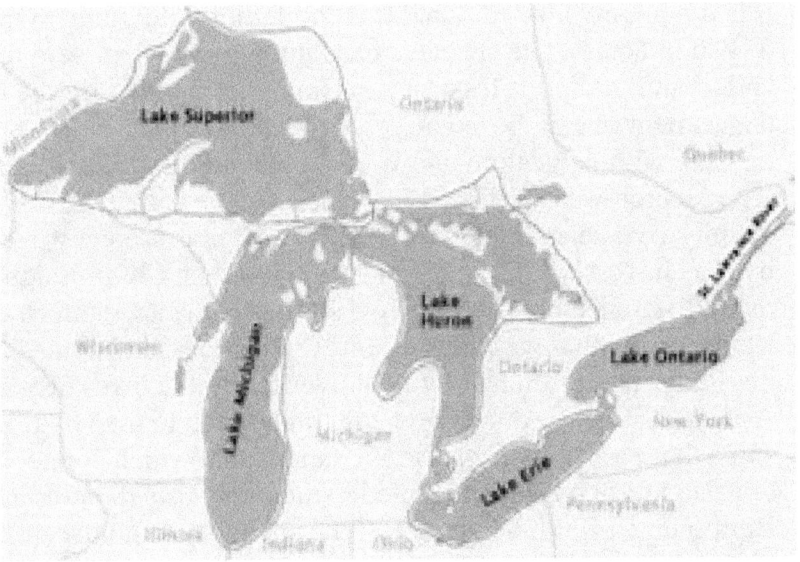

Lake Erie, being the shallowest of the Great Lakes and having its runoff, more or less, blocked by the retaining structure of Niagara Falls is the most polluted of the lakes. It receives pollution from the upper three and its basin can only pass on a fraction of it to Lake Ontario which flows uninhibited to the Atlantic Ocean. Some of the Lake Erie water is diverted to the Welland Canal system but this is top water and does not affect the concentration of "heavy metals" in the deeper waters of the lake. Seasonal thermal shifts do not carry the pollutants upward enough to dispel them from the lake. So it would seem evident that the species with the least tolerance to pollution would be found in Lake Superior and those with high tolerance would be found in Lake Erie.

Photos that were provided by fish commissions and internet sources are labelled according to the state of origin,

This work

This work details the species of fish known to be in the Great Lakes. It is not about "how to fish." It is more of a field guide to the species of fish in lakes. The species highlighted in this text were researched and identified by Professor Paul Thomas a member of the biology department of Edinboro University of Pennsylvania. He worked at the task with dedication and without the aid of state or federal grants. His research was published in *Fishes of Erie County* in 1993. Twenty-five years later the results of a similar project was published by the University of Toledo and the Ohio Fish Commission. The main difference between the two studies is the appearance of the Round Goby that was not present when Professor Thomas completed his study. The Round Goby is indigenous to the lakes of western Asia and is believed to have been transported to the Great Lakes in ships that passed through the Welland Canal which connects Lake Ontario to Lake Erie. The Toledo study also lists the presence of the Bigmouth Buffalo, a member of the sucker-type family of fish. This difference also illustrates the danger of using common names instead of scientific names in identification. Local terms such as bighead and bigmouth as well as sucker or minnow are useful but not acceptable in the realm of science. This book should satisfy the desire to know the fish of the Great Lakes.

The illustrators for the work by Professor Thomas were Jerry McWilliams and Adah Ellis-Anderson and the pen and ink illustrations in this work are by those two naturalist artists. For other information on who assisted Professor Thomas one may refer to his work *Fishes of Erie County.*

Fish commissions, of course, keep tabs on the species of fish and their welfare and they take measures to ensure that the desirable species continue. A significant listing of the fish in Lake Superior was presented in 2014 and is another basis for the information in this work. This list appears near the end of this work. It contains the common name of the species, as well as the scientific names. Common names are useful but not in the realm of science. For instance the Superior study refers to the Common Cisco while the Thomas study regards the same fish as Lake Herring.

Author's Notes. Part I

Except for a disputed small portion of land in Wisconsin the area and drainage basin of the Great Lakes was covered by a mile thick layer of glacial ice. In 1962 I began doing research for my book *The Geology and Geography of Erie County* which eventually was published by the Pennsylvania Geological Survey.

    Part of my field research involved walking from the crest of the plateau south of the lake to the lake. The terrain consisted of a slope, then a flat area, then a slope, then a flat area. There were four distinct flat areas in the land that I traversed before I reached the lake. These flat areas were the former beaches of the lake that existed as the ice sheet melted and trapped glacial melt water between the slope and the ice sheet.

    The glacial melt water of all the Great Lakes area flowed in many directions according to geologic investigations and eventually joined into an east flowing river emptying into the Atlantic Ocean. Various periods of natural dams and erosion have made the lakes what they are today.

    As my research continued I began to wonder how the flora and fauna of the lakes began to accumulate. The advance of vegetation seemed easy enough to explain with spores and seeds carried on the breeze or by slow migration over many years from south to north. But, what about the fish? How did ocean species get over Niagara Falls? That's a pretty high jump, even for a salmon. There's a lot of room for speculation here.

Then eventually there came to my mind the question of native species and invasive species. Generally, a native species is one that existed when people started recording them which is when the Europeans, an invasive species, entered the Great Lakes Region. An invasive fish species is one that entered after the initial recording.

Eventually, my travels on and around the Great Lakes indicated there were not only physical differences in the depths and shapes of the lakes but also difference in water quality, temperature and movement, not only of water but the inhabitants of that water.

The lakes support over ten million people whose farm and city lifestyles affect the lakes. However, that is another topic and this work is about the species of fish in the lakes and some system of classification in order to understand more about nature and our world. The universe is also another topic that's difficult to comprehend.

Authors Note. Part II

I have traveled on and around the Great Lakes many times in my ninety years of existence. I and my family and companions have camped on the shores and fished in the waters of all the lakes. But, since I live on the slope overlooking Lake Erie that is where I have spent most of my time.

I have been fishing in Lake Erie and its tributary streams, off and on, for just under seventy years. There were winter days of ice fishing and hot summer days of trolling a one pound sucker for many hours. We also did a lot of night fishing in the shallow areas around the lake edges.

One day I caught a fish that was a stranger to me and my companions. After much asking around it turned out to be a *Burbot*, a type of codfish. It made me wonder what all was in the lake but my occupation and other pressing problems did not permit me to engage in research of that nature. The Fish Commission seemed to be mostly concerned with the commercial aspects of its mission. What were all those little fish swimming close to shore? Did they have a name? If something does not have a name it is difficult to have a discussion about it.

My first fishing experience in the lake was in 1951 and Blue Pike were plentiful. The Blue Pike are now gone. There were no Sea Lampreys in the lake at that time as well as an absence of other species that are plentiful today. That is something to think about when you contemplate life on the planet and the changing environment and its effect on plants and animals. The hot topic of the moment is "global warming." Will the water of the lakes warm to the point where the cold water species disappear and alligators migrate north from the Everglades and South Carolina.

When I first started fishing in Lake Erie one of my older companions was an octogenarian Mr. Matthews. We were about a hundred yards off shore when he took his thermos bottle cap off the thermos and scooped up a cupful of water and drank it. He wouldn't dare do that today. The industrial growth around the lake about twenty years after World War II began to contaminate the lake with heavy metals. Sewage has always been a problem and still is. At one time there was much talk about the death of the great lakes. The river running through Cleveland was so polluted at one time the surface was set on fire and the fire department responded in order to save a wooden bridge over the water.

The Great Lakes are simply a widened river system emptying into the Atlantic Ocean. River systems eventually flush out, but areas such as Niagara Falls keep the heavier pollutants in the lake basin.

The passage of the clean water, air and soil acts has helped to bring the lake back to acceptable levels of purity, but there is still a long way to go.

When I first started fishing in Lake Erie there were sixteen commercial fishing companies operating around the city of Erie. Lake Erie Perch was on every menu in every restaurant. Today, there are no commercial fisheries in Erie County. Lake Erie perch can still be found on some menus in Mom and Pop restaurants.

The Fish Commissions and state legislatures have decided to emphasize sport and recreational fishing in the lake rather than commercial fishing.

The fish commissions and health departments of the states bordering on Lake Erie have all issued warnings about eating fish from the lake. They recommend eating only one meal of fish a month and pregnant women should not eat fish from the lake under any circumstance.

There are some methods of cleaning and cooking fish that are less formidable than others. For instance, only eating the slab sides of fish caught in open waters would be less hazardous than eating the entire fish obtained from shallow muddy bays.

The lakes are slowly recovering from the excesses of pollution in the last century. With proper legislation, and enforcement of regulations, the lakes will further recover. It is important in many ways. The water in my home comes from Lake Erie and I often wonder about it as I brew a pot of morning tea.

The original books by **Professor Paul Thomas** were designed to be used for accurate field identification and promoted scientific criteria such as jaws, teeth, rays, and fins. Over the years he identified 112 different species in the lake. His research was published in *"Fishes of Erie County,"* published by Allegheny Press in 1993. It was a small printing and not many copies of the work can still be found.

This work is designed more for the person who might happen to catch a strange fish and wonder what it is. Of course, scientific data must be retained and along with it, scientific names. True scientists are alarmed at a description of a species without a scientific name attached to it. The list of Professor Thomas can be found at the end of this work and it consists of common names and their scientific identification. It is a good list to consult since it lists the species present at a certain time in history, in this case 1992.

There is also an updated list provided by the University of Toledo in 2017. Some species in the Toledo list were not in the lake at the time Professor Thomas did his study. Some fish that are obviously in the lake at this time were not included in the Toledo data because they weren't found in the time period of the survey. The muskellunge is not in the list because it was not detected in the selected area during the time of the study, but it was found in other parts of the lake during that time period.

Everyone knows what a catfish looks like so there is no need to dwell on scientific names and descriptions. We might point out the slight differences between related species but don't dwell on it. Having the fish in your possession and looking at the generic illustration should be adequate for identification.

The pen and ink illustrations are from *Fishes of Erie County* by Paul Thomas. Pen and ink illustrations are much better than photographs for identification once you begin to use them. For instance, lip striations may be in the illustration of a sucker but not readily visible in a photograph.

Invaders

The Welland Canal connecting Lake Ontario to Lake Erie for shipping was begun in 1830 and that was the year of the first recording of the Sea Lamprey in Lake Ontario. The canal became a gateway for ocean species to work their way inland.
The lamprey appeared in Lake Erie in 1936. A few years later their population exploded and there was a drastic decline in the population of lake trout, one of the most desirable fish for human consumption. The lamprey then moved north and west inhabiting the other lakes. Measures to combat the lamprey were immediately taken with some success because the lamprey moved into the fresh water streams to spawn.

The next upheaval in the Great Lakes was the arrival of the Alewife in 1949. Their high reproductive rate caused a decline in other fish populations by causing a decline in available food. This disruption of the environment contributed to the decline of commercial and sport fishing at that time.

Vast numbers of Alewife die off over the winter months and their rotting carcasses, along with their cousin the gizzard shad, line the shores of the lakes in the past as they still do to this to this day every spring after the ice melt.

Once control programs for the lamprey and alewife were somewhat successful a program of fish stocking began in earnest. Planting of lake trout was a high priority. Lake Trout planting began in 1965. The program was not wildly successful but it did have some positive results. Coho salmon were stocked in 1966 and Chinook salmon followed the next year. The success of the salmon stocking led to the stocking of other species which included Rainbow Trout, Brown Trout, and Muskellunge which were not depleted but fish and wildlife commissions could see an economic benefit with an increase in their numbers. Many private fish organizations also participated in stocking in those years as they do today.

An eight inch alewife.

A sixteen inch gizzard shad.

II. Fish Families

Broad Classification of the fish in the Great Lakes

Some terms used in the following classifications.

adipose fin – fleshy, rayless structure on the mid-dorsal line between the dorsal and caudal fins.
anal fin – mid ventral fin, posterior to the anus.
barbel – icicle shaped projection or fleshy flap, usually on the chin, lips, or nose.
buccal funnel – circular mouth cavity in lampreys
caudal fin – tail fin
concave – arched inward
convex – arched outward
ctenoid scales – have a comb-like posterior margin

cycloid scales – have smooth posterior margin and are found on soft ray fishes
dorsal – back or upper portion of the fish
dorsal fin – median fin on the back
emarginate – notched but not forked tail
fin rays – soft rays which are usually branched and flexible and spiny rays.
gill rakers – knob-like projections on the anterior edge of the first gill arch.
gular plate – a body plate between the lower jaws of a bowfin
heterocercal – caudal fin with vertebrate column extending into the upper lobs, usually making it longer than the lower lobe.
homocercal – caudal fin with lobes of equal size
keel – midline of the body
mandible – lower jaw
maxilla – bones of the upper jaw
opercle – bony covering of the gills

paired fins – ventral fins, pectoral (anterior) and pelvic (posterior)
papilla – fleshy projections
parr marks – oblong blotches along the sides of young trout.
pectoral – breast area
pelvic – abdominal area
posterior – toward the back
serrated – toothed
spiny rayed fishes - spiny rays anterior to soft rays in dorsal and ventral fins
spiracle – opening above and behind the eye
striated – streaked
teardrop – a dark spot below the eye
terminal – extending to the end
ventral – belly or lower part of the fish
vomer – bone in front of the roof of the mouth.

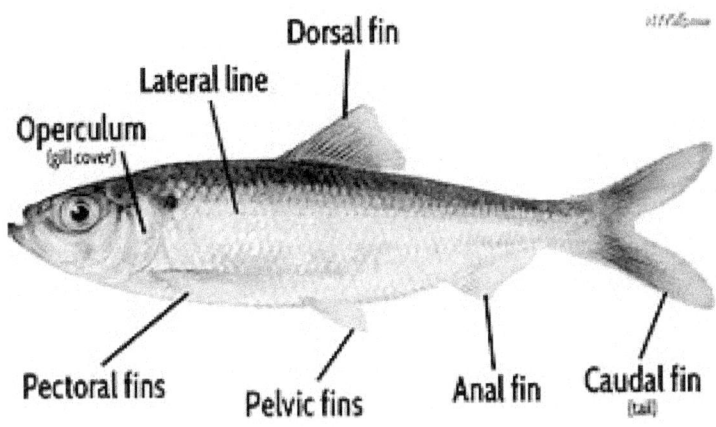

Field characteristics of the different families of fish in the Great Lakes. This is the Family classification system of Paul Thomas, and the fish he found in those families. The scientific names of the species are listed under the several indexes in the last pages of this work. His data was published in 1993

1.
Lampreys – *Petromyzontidae* - eel shaped body, no true jaws, circular funnel shaped mouth, scaleless, no paired fins, long dorsal fin, more or less continuous with the caudal fin, seven pairs of external gill openings, and a single median nostril.

     Examples found: Sea lamprey, Ohio lamprey, Northern Book lamprey, Mountain Brook lamprey, Silver lamprey, American Brook lamprey

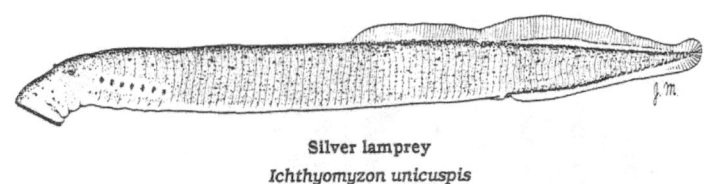

Silver lamprey
Ichthyomyzon unicuspis

The parasitic Silver lamprey has one dorsal fin. Its average adult size is around 15 inches.

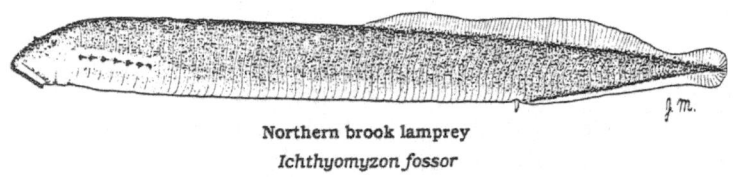

Northern brook lamprey
Ichthyomyzon fossor

The Northern brook lamprey is not parasitic. It has a poorly developed funnel mouth. It has one dorsal fin. It only grows to around six inches. Its body is very soft to the touch. Compare its mouth to the parasitic mouth of the Silver Lamprey above. It feeds on insects and algae found on rocks.

2
Sturgeons – *Acipenseridae* - heterocercal tail, subterminal mouth preceded by two pairs of barbels under a conical snout, large bony plates on the head and in horizontal rows on the body, spiracle present in Great Lakes forms.
Found: Lake Sturgeon

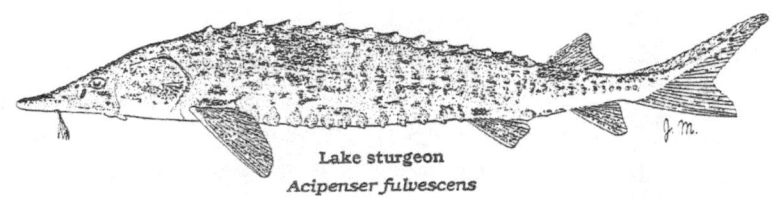

Lake sturgeon
Acipenser fulvescens

The Lake sturgeon has a heterocercal tail. Its subterminal mouth is preceded by 2 pairs of barbels. It has bony plates on its head as well as in horizontal rows on its body. It is a protected species. It can grow to five feet in length.

3
Gars – *Lepisosteidae* – abbreviated heterocercal tail, caudal fin rounded, snout extended into beak, teeth long and sharp and conical, thick diamond shaped ganoid scales, dorsal fin shorter and near the caudal fin, and a cylindrical body.
   Found: Longnose gar, Spotted gar

Longnose gar

The Longnose gar has an abbreviated heterocercal tail and a rounded caudal fin. Its snout is extended to a beak. Its teeth are long and sharp. It can grow to three feet in length.

4
Bowfins - *Amiidae* – abbreviate heteroceral tail, caudal fin rounded, cycloid scales, head covered with bony plates, bony plate fills space between lower jaws, long dorsal fin runs over most of the length of the back, the back and sides are mottled, spot at base of caudal fin is most prominent in adult males.
    Found: Bowfin

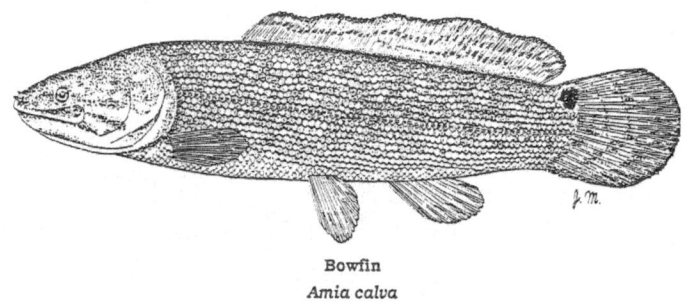

Bowfin
Amia calva

The bowfin maxes out at about 25 inches.

5
Herrings - *Clupeidae* – ventral midline of the belly is strongly serrate, head scaleless, soft-rayed, pelvic fins abdominal, eyes partly covered with eyelids, body slab sided, no lateral line.
Found: Alewife, Gizzard shad

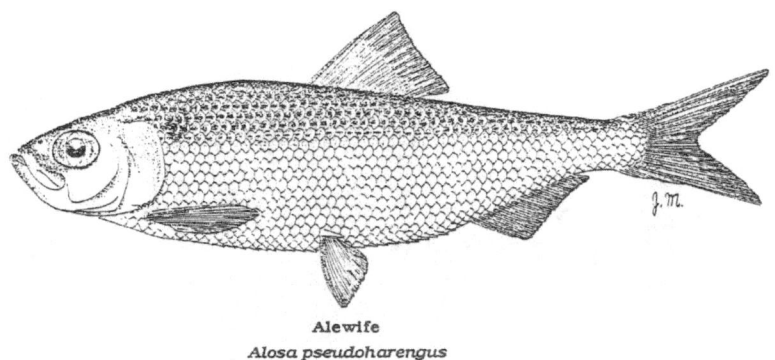

Alewife
Alosa pseudoharengus

The Alewife has a mouth that opens upward and a dark shoulder spot on its body behind the opercle. The adult length may reach twelve inches, but rarely does.

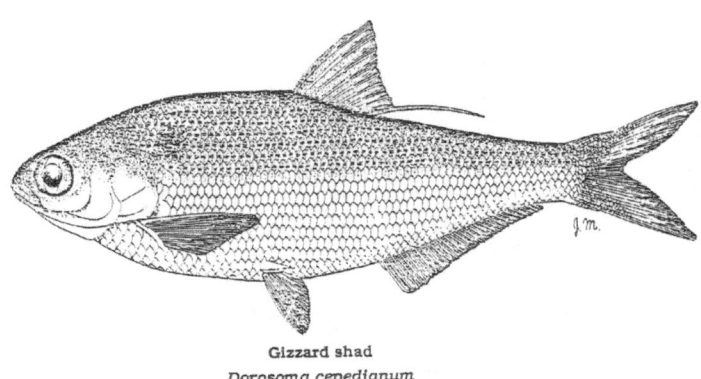

Gizzard shad
Dorosoma cepedianum

The Gizzard shad mouth is overhung by a rounded snout. It has a dark spot on its shoulder behind the opercle and above its pectoral fin. It grows to just short of twelve inches.

6
Trouts – *Salmonidae* - head scaleless, soft rayed, adipose fin, auxiliary process present at the base of the pelvic fin, pelvic fins are abdominal, Found: Lake Herring, Lake Whitefish, Coho Salmon, Rainbow Trout, Sockeye Salmon, Chinook Salmon, Brown Trout, Brook Trout, Lake Trout.

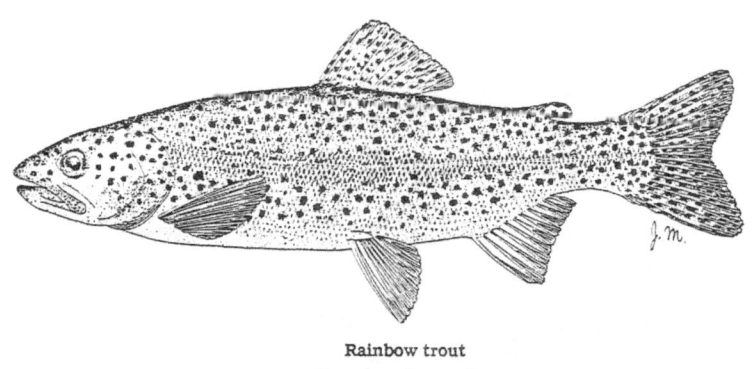

Rainbow trout
Oncorhynchus mykus

The Rainbow trout, aka Steelhead, has sides with rainbow hues. There are no lateral reddish spots. Adults in the lakes can reach two feet in length.

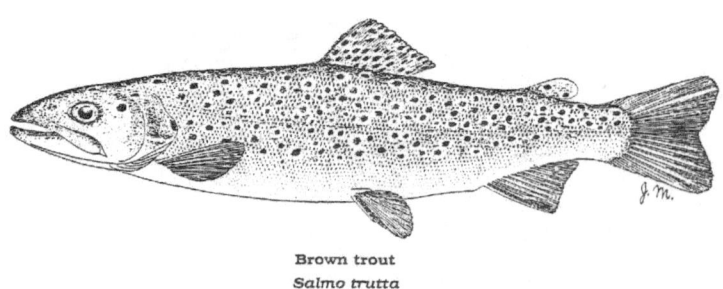

Brown trout
Salmo trutta

The Brown trout has lateral red spots. It has no white leading edges on its paired anal, or caudal fins as does the brook trout. The brown trout can grow to two feet in length.

Chinook salmon
Oncorhynchus tschawytscha

The Chinook salmon has a sharp snout with teeth. All of its fins have a few black spots with a lot on its dorsal lobe. It is also known as the King salmon.

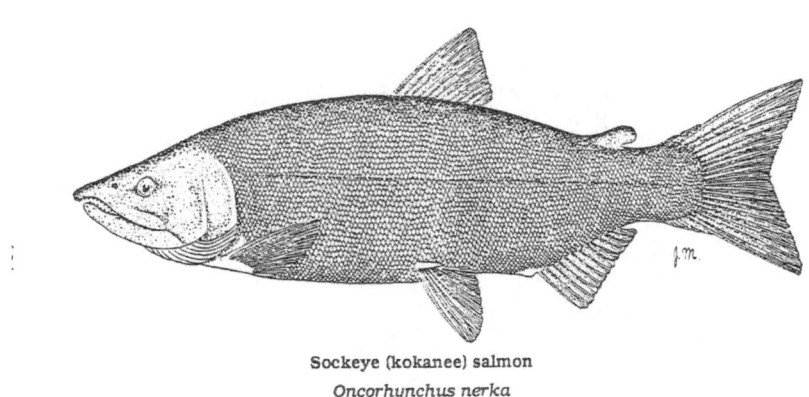

Sockeye (kokanee) salmon
Oncorhynchus nerka

The Sockeye salmon has no black spots on the back of its caudal fin. It is bluish above and silvery below. It grows to about 24 inches. It is not plentiful in the lakes and has not been identified as living in Lakes Superior and Michigan.

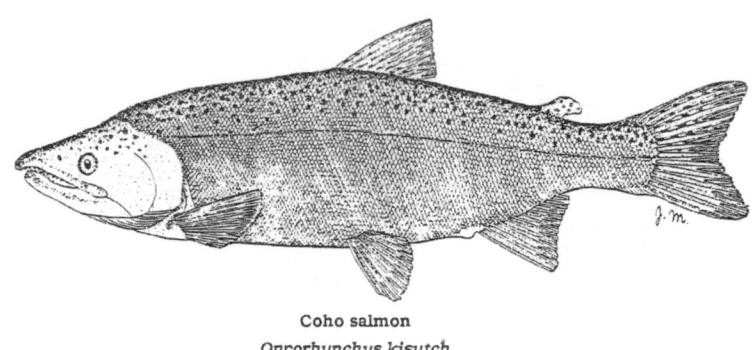

Coho salmon
Oncorhynchus kisutch

The Coho salmon has a sharp snout with teeth. It has small black spots on its sides above the lateral line. It grows to about two feet. It is also known as the Silver salmon.

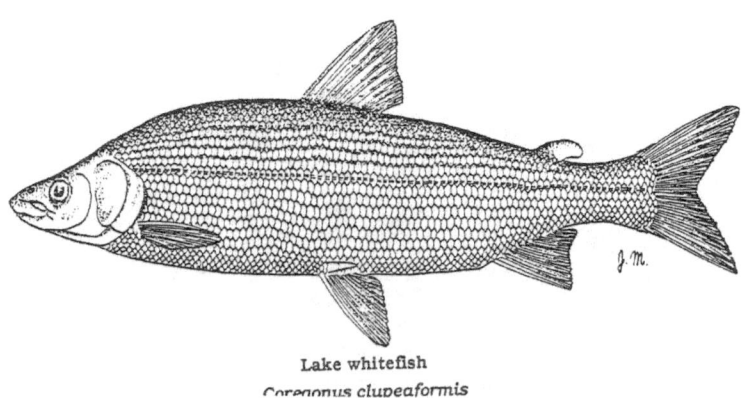

Lake whitefish
Coregonus clupeaformis

The Lake whitefish has a blunt or rounded snout and a small subterminal mouth. It grows to about twenty inches.

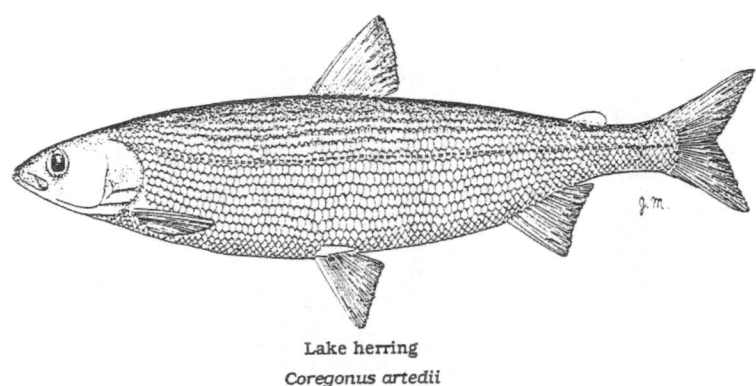

Lake herring
Coregonus artedii

The Lake herring has a sharp snout and is all but toothless. It can grow to just short of twenty inches. It is sometimes referred to as the Cisco.

7
Smelts – *Osmoridae* – stout teeth on vomer, head scaleless, soft rayed, adipose fin, pelvic fins abdominal, more or less round in cross section, silvery coloring. Found: Rainbow smelt.

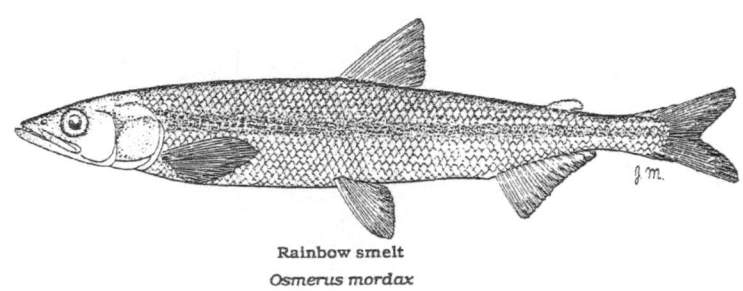

Rainbow smelt
Osmerus mordax

The Rainbow smelt has a large mouth, well developed teeth and a cigar shaped body. It can grow to 15 inches in the lakes.

8

Mudminnows – *Umbridae* - jaws with teeth, cycloid scales, head with scales, soft rayed, dorsal fin rather far back, pelvic fins abdominal, dark vertical bar at the base of the tail, rounded tail. Found: Central mudminnow

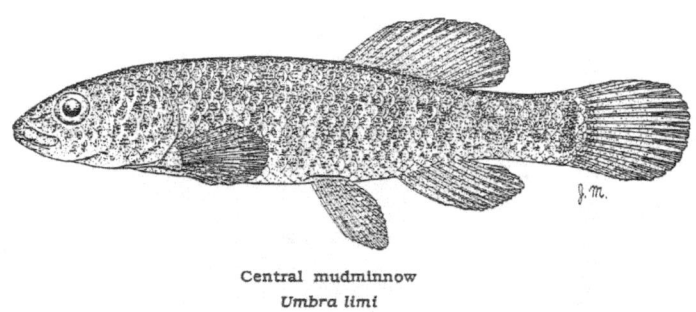

Central mudminnow
Umbra limi

The Central mudminnow is found in calm clear water. It spawns in the spring. The illustration above is the actual size of most adults – four inches or less.

9

Pikes – *Esocidae* – top of head shaped like a duck bill, stout sharp teeth in jaws, cycloid scales, head with scales, soft rayed, dorsal fin far back on the body, pelvic fins abdominal, body long and cylindrical. Found: Northern Pike, Grass Pickerel, Muskellunge, Tiger Muskellunge, Chain Pickerel

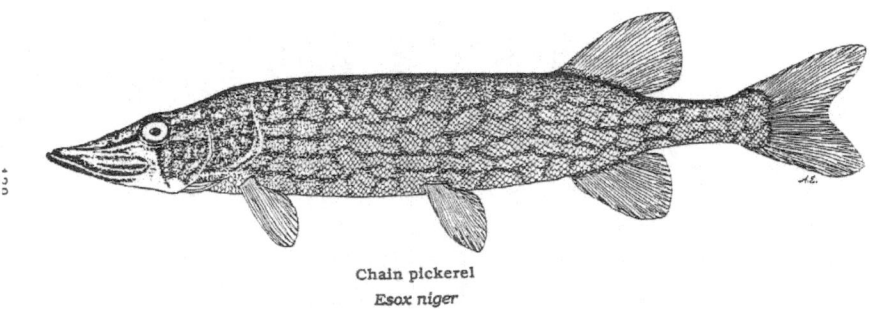

Chain pickerel
Esox niger

The Chain pickerel does not appear in the Toledo or Minnesota lists. Its cheeks and opercles are fully scaled. It is rare in the lakes. It can grow to two feet.

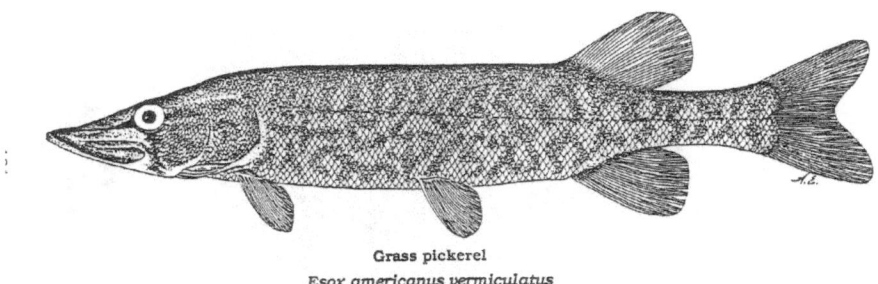

Grass pickerel
Esox americanus vermiculatus

The grass pike is also not on the Toledo or Minnesota lists. Its cheeks are fully scaled. It is the smallest of the pikes and grows to less than twenty inches.

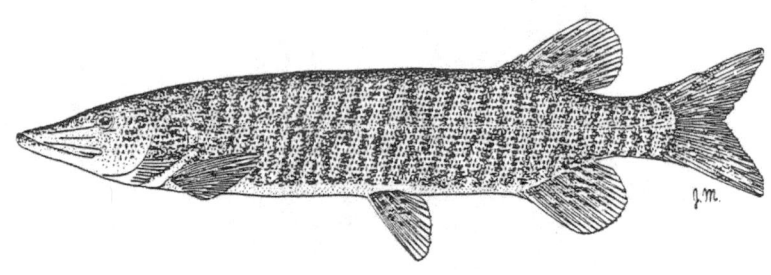

Tiger muskellunge
Esox masquinongy/Esox lucius
(hybrid)

The *Tiger muskellunge is a hybrid fish developed by genetic manipulation and cross breeding. Its body is plumper than the typical Muskie. It can grow to more than two feet.*

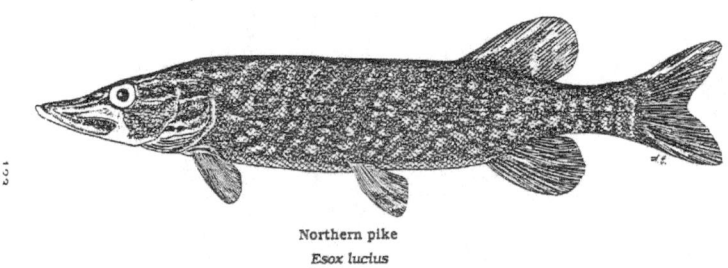

Northern pike
Esox lucius

The *Northern pike has fully scaled cheeks and has light spots on a dark background. It grows to around thirty inches.*

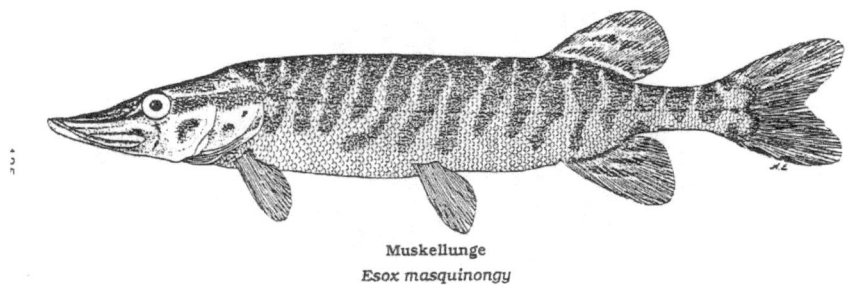

Muskellunge
Esox masquinongy

The Muskellunge is the premier sport fish. Its cheeks are scaleless on the lower half. It has dark spots or bars on a light background. Some Muskies have no markings. Adults average around 40 inches but some record fish have reached 60 inches.

10

Minnows – *Cyprinidae* – toothless jaws, cycloid scales, scaleless head, soft rayed, single dorsal fin, pelvic fins abdominal. **Found:** Bluntnose minnow, Central Stoneroller, Goldfish, Redside dace, Common dace, Silverjaw Minnow, Tonguetied minnow, Bigeye chub, Streamline chub, Silver chub, Horneyhead chub, River chub, Golden shiner, Emerald shiner, Striped shiner, Blackchin shiner, Common shiner, Spottail shiner, Silver shiner, Rosyface shiner, Spotfin shiner, Sand shiner, Redfin shiner, Mimic shiner, Southern Redbelly dace, Bluntnose minnow, Fathead minnow, Blacknose dace, Longnose dace, Creek chub, Pearl dace.

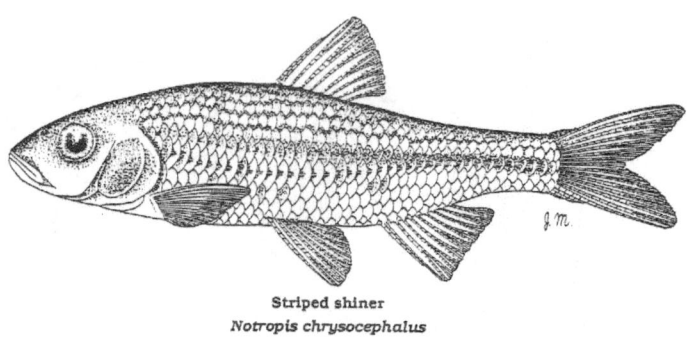

Striped shiner
Notropis chrysocephalus

The Striped shiner has large eyes and dark lines on the back forming Vs when viewed from above. It can grow to seven inches.

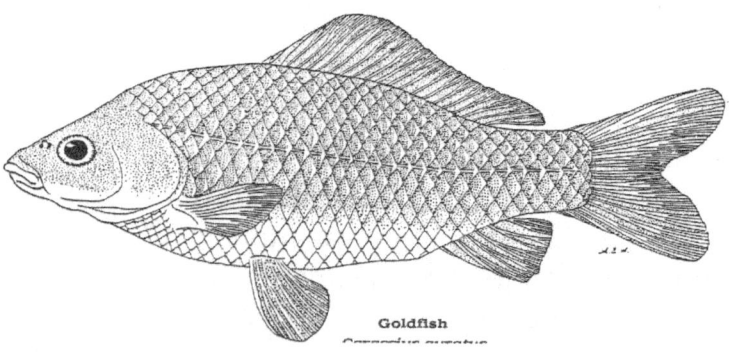

Goldfish

The Goldfish is not a native species. It does not have barbels but it does have a long dorsal fin. I caught one that was 12 inches long in Lake Erie and one that was eight inches long in Lake Huron.

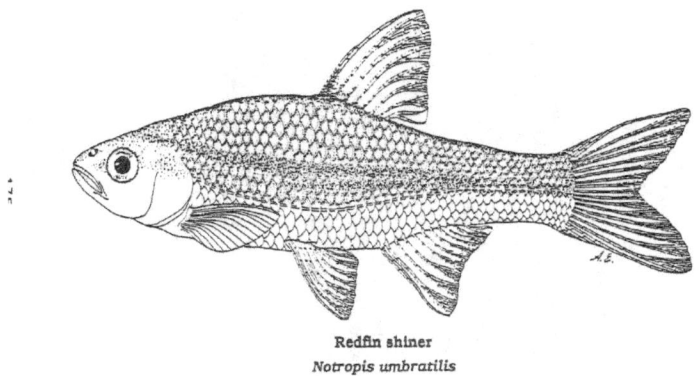

Redfin shiner
Notropis umbratilis

The Redfin shiner has the red color on its fin when it is a breeding adult. Its body is bluish silvery. Its mouth is large and at an oblique angle. Its upper lip is separated from the jaw by a groove. It only grows to about three inches.

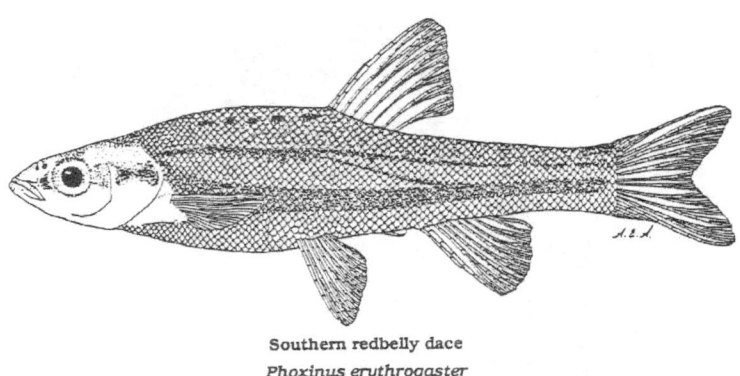

Southern redbelly dace
Phoxinus erythrogaster

The Southern redbelly dace has very small body scales. Its mouth is at a slight oblique angle. It features two dusky lateral bands. It grows to three inches max.

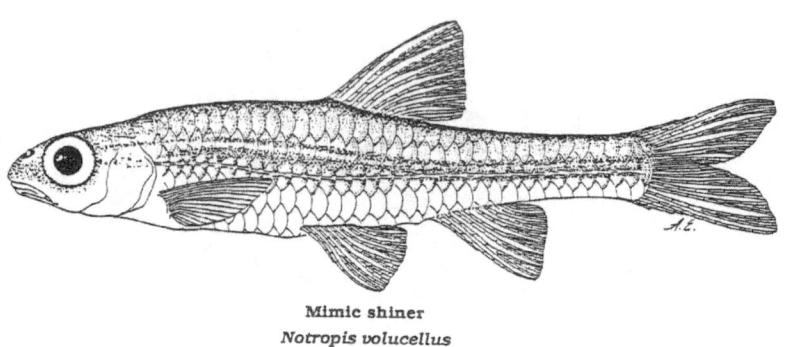

Mimic shiner
Notropis volucellus

The Mimic shiner and Sand shiner are similar, but this one has a pigmented anus. It grows to slightly less than three inches.

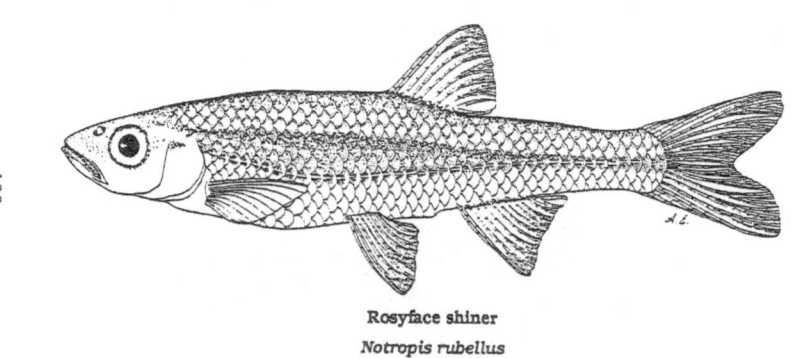

Rosyface shiner
Notropis rubellus

The Rosyface shiner has a sharp snout and a large mouth. Its lateral line is complete but curved in the front. It grows to about three inches.

Spotfin shiner
Notropis spilopterus

The Spotfin shiner has a small eye compared to the Rosyface shiner. Its mouth is at an oblique angle. It has a distinct lateral band. It grows to about four inches when mature.

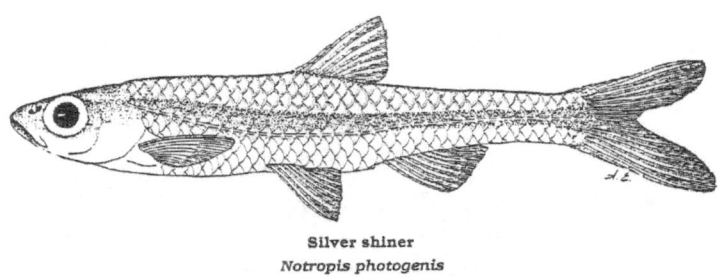

Silver shiner
Notropis photogenis

The Silver shiner has large eyes and a silvery body. It has a dark mid-dorsal stripe. It has two dark crescent-shaped spots between its nostrils. The front of its dorsal fin is slightly behind the front of its pelvic fins. It grows to slightly more than four inches when fully mature.

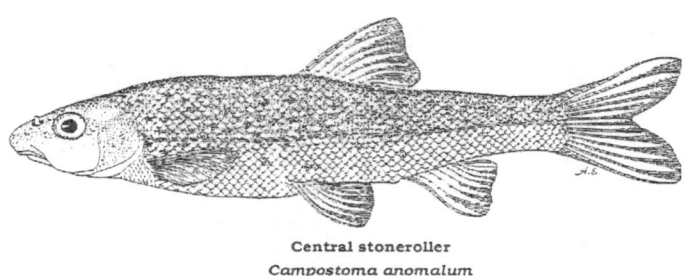

Central stoneroller
Campostoma anomalum

The Central stoneroller snout overhangs the mouth. It has a cartilage ridge on its lower jaw. The upper lip is separated from the jaw by a groove. It maxes out at around six inches.

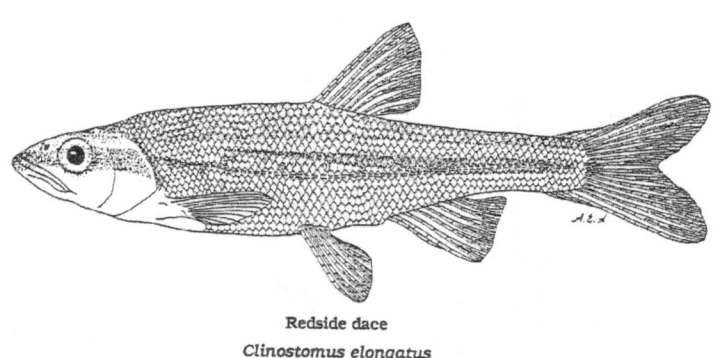

Redside dace
Clinostomus elongatus

The Redside dace has reddish color on its sides and a sharp snout. Its jaw goes back to its eye. The adult gets to just short of three inches.

Silverjaw minnow
Ericymba buccata

The Silverjaw minnow has visible bones on its mandible, maxillary, and sides of its lower head. The upper jaw is about as long as its eye diameter. It grows to about three inches.

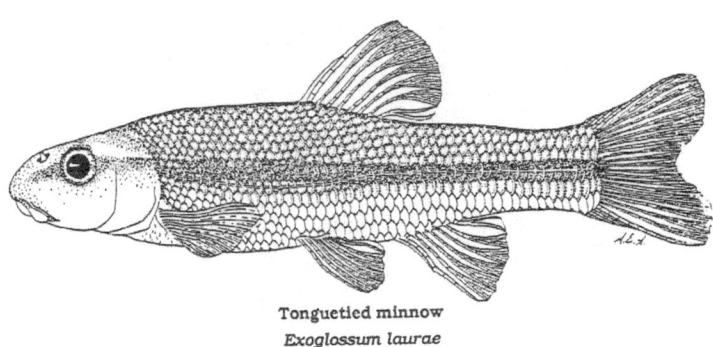

Tonguetied minnow
Exoglossum laurae

The Tonguetied minnow usually has a small barbell present. Its mouth is subterminal. Its upper lip is connected to the jaw by a central bridge of flesh. Its lower jaw appears lipless. It grows to about six inches.

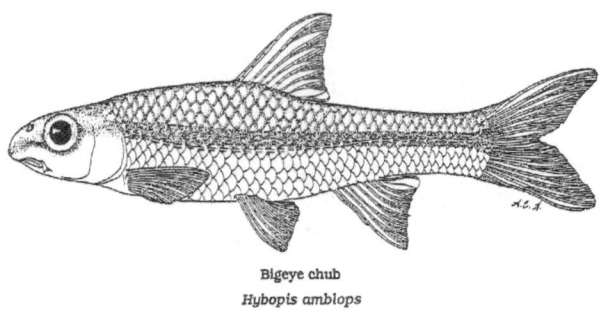

Bigeye chub
Hybopis amblops

The Bigeye chub has a large eye and a snout that overhangs its mouth. It has a barbel at the back end of its upper jaw and a dark lateral band. The front of the dorsal fin is over the front of the leading edge of its pelvic fins. It grows to less than four inches.

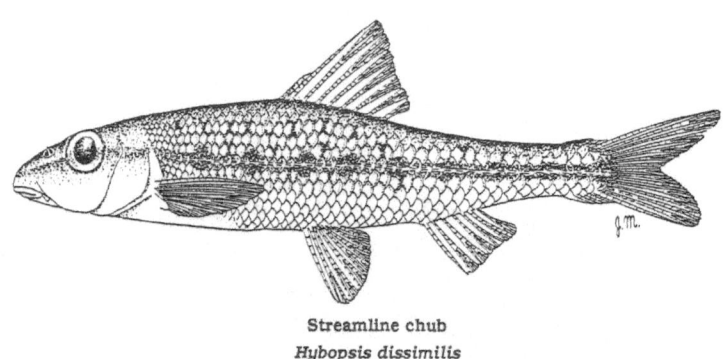

Streamline chub
Hybopsis dissimilis

The Streamline chub has a large eye and a snout that overhangs its small mouth. Its lips are thick and it has a barbel at the back end of its upper jaw. Its body has black spots or blotches. Its back has X or W marks on it. It grows to four inches.

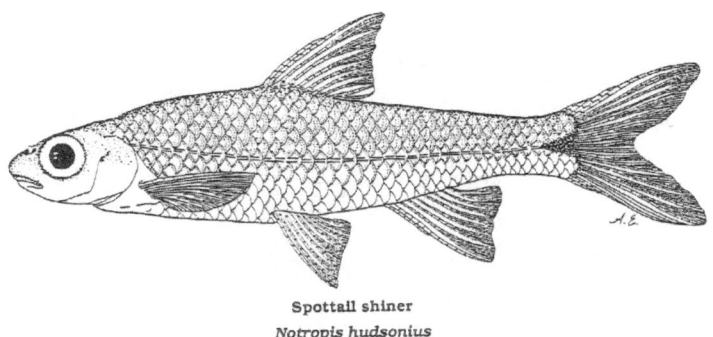

Spottail shiner
Notropis hudsonius

The Spottail shiner has a well-defined black spot at the base of its tail. Its dorsal fin is more anterior than for most other shiners. It grows to about five inches.

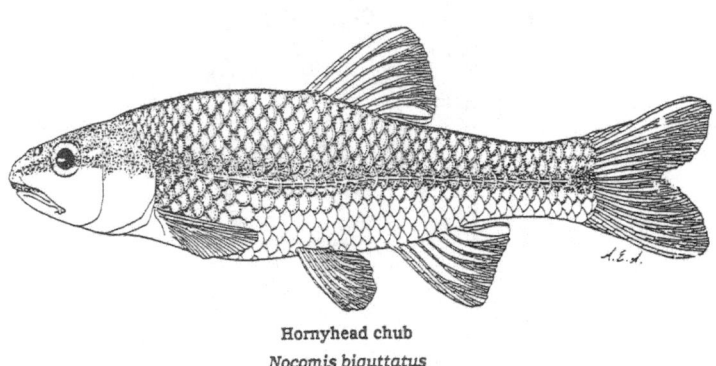

Hornyhead chub
Nocomis biguttatus

The Hornyhead chub has a dark round spot at the base of its tail. It has a very small barbel at the posterior end of its upper jaw. Its caudal fin is often orange, especially in the immature. It grows to nine inches.

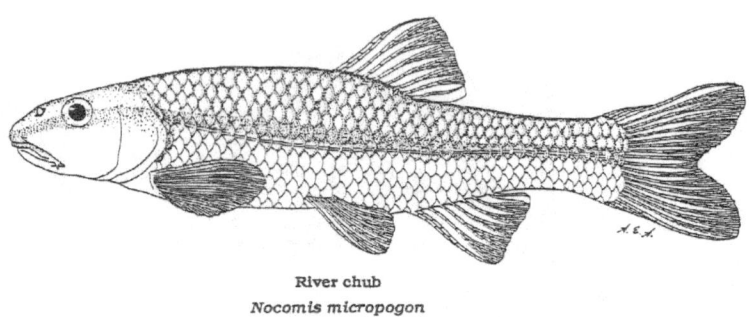

River chub
Nocomis micropogon

The River chub has an almost obscure caudal spot. Its snout slightly overhangs a large oblique mouth. The adult may grow to just short of nine inches.

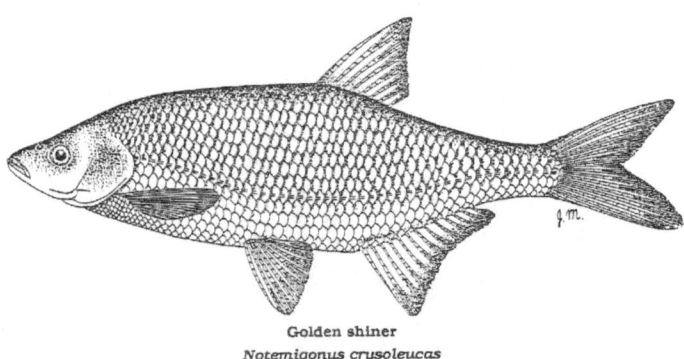

Golden shiner
Notemigonus crysoleucas

The adult Golden shiner is golden. It has a small oblique mouth and a strongly decurved lateral line. It has a long anal fin base. Its body is deep and slab sided. The adult grows to seven inches.

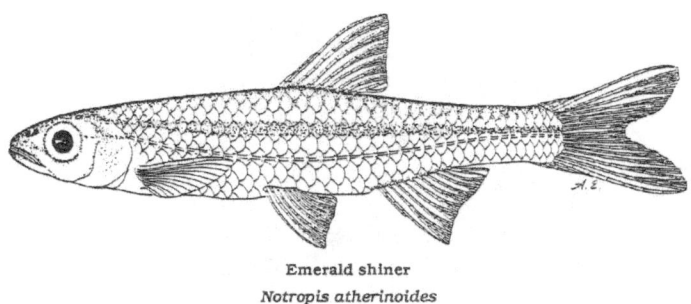

Emerald shiner
Notropis atherinoides

The Emerald shiner has a silvery body with emerald reflections. It has a faint lateral band and its body is slab-sided. It grows to just short of four inches. It is the most sought minnow for use as a bait fish and brings a premium price for that purpose.

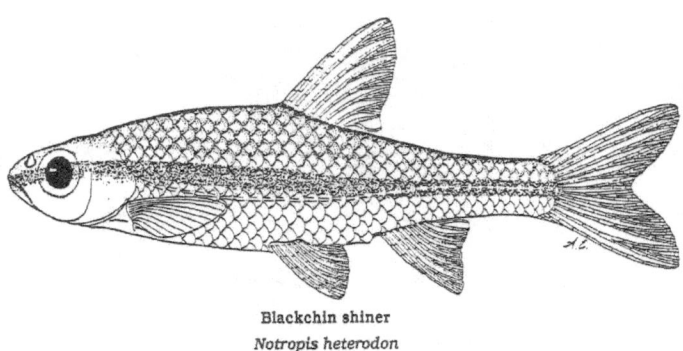

Blackchin shiner
Notropis heterodon

The Blackchin shiner has a black snout and a black chin. Its lateral band extends through the eye to the chin. Its lateral line is often incomplete. It grows to slightly less than three inches.

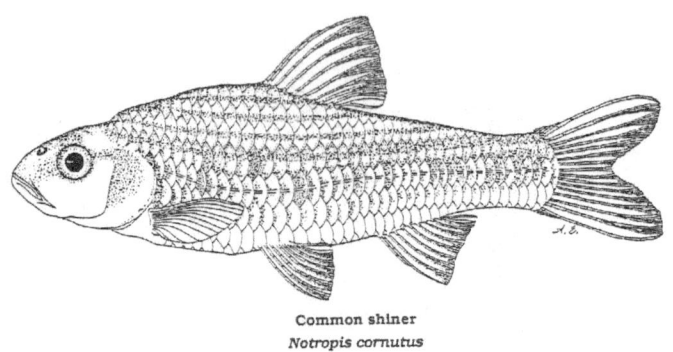

Common shiner
Notropis cornutus

The Common shiner has large elevated diamond shaped scales on its sides. It has a prominent mid-dorsal stripe and no lateral bands. It grows to about 8 inches.

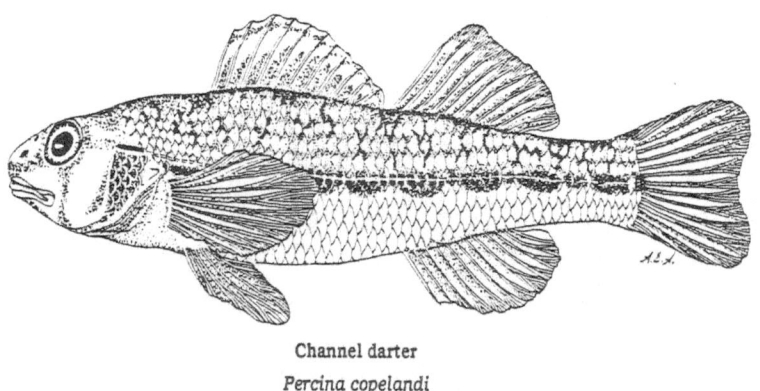

Channel darter
Percina copelandi

The Channel darter has a row of small blotches on its sides. Its snout is blunt or rounded. Its midline on the belly is naked or with specialized scales. It grows to less than three inches.

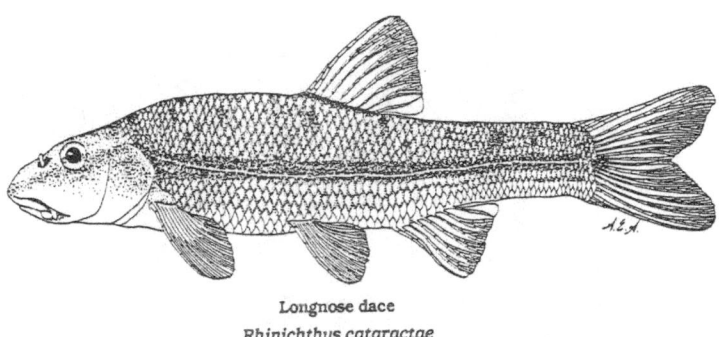

Longnose dace
Rhinichthys cataractae

The Longnose dace snout projects far beyond its horizontal mouth. Its eyes are superolateral. It has small scales and its back and sides are mottled. It grows to just short of five inches.

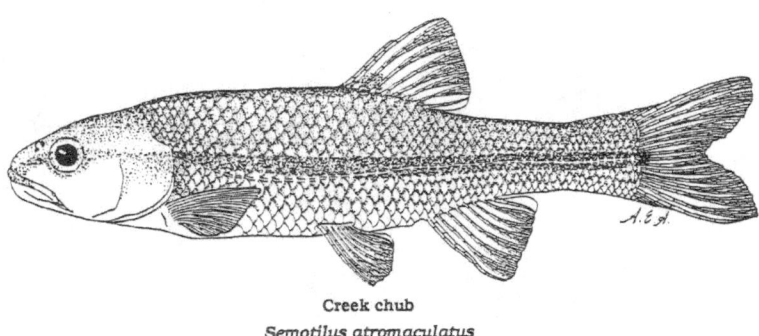

Creek chub
Semotilus atromaculatus

The Creek chub has a spot at the anterior base of the dorsal fin. It has a dark spot at the base of its tail. It has a large mouth and a very small barbel in advance of the posterior end of its upper jaw. This is absent in the young of the species. Its lateral band extends through to the eye. It may grow to eight inches.

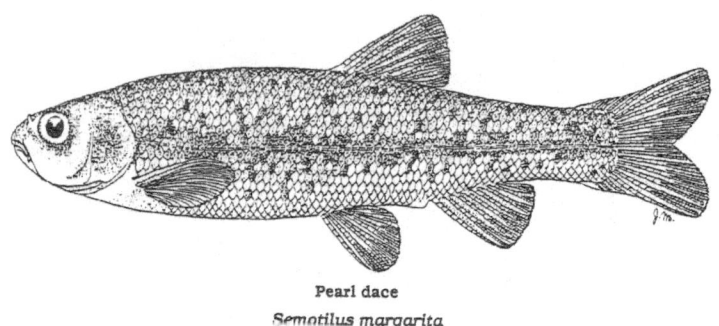

Pearl dace
Semotilus margarita

The Pearl dace had mottled back and sides. It grows to about four inches.

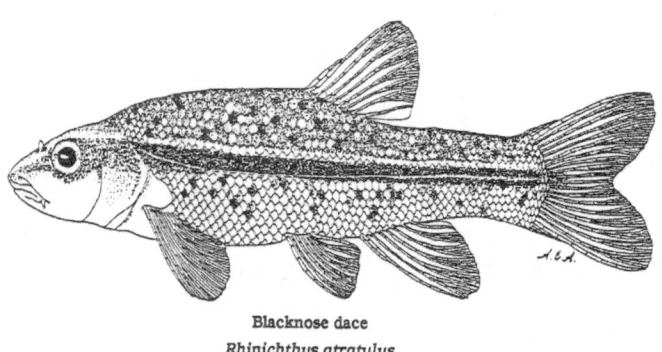

Blacknose dace
Rhinichthys atratulus

The Blacknose dace does have a black nose. Its lateral band extends through the eye to the nose. It has silvery sides and its snout overhangs slightly the mouth. Its back and sides are mottled. It grows to just short of four inches.

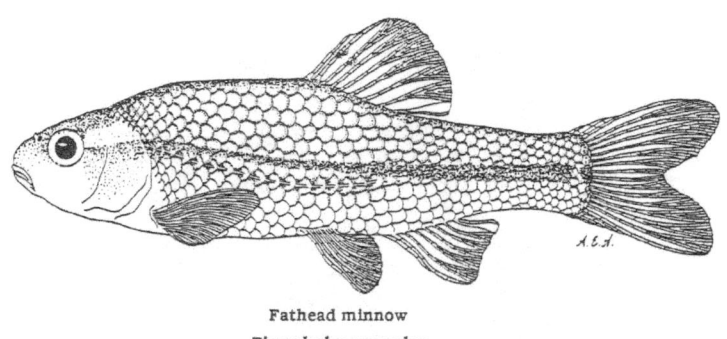

Fathead minnow
Pimephales promelas

The Fathead minnow has scales crowded toward the head. Its mouth is oblique. Adults have a horizontal dark bar across its dorsal fin, about half way up. Its lateral line is often incomplete. It grows to slightly less than three inches.

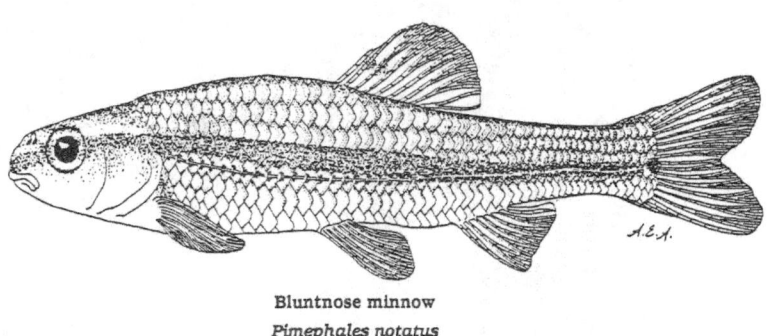

Bluntnose minnow
Pimephales notatus

The Bluntnose minnow does have a blunt nose and its snout slightly overhangs its small mouth. Its lateral line is wide. It has black spots at the base of the front of its dorsal fin and a black spot at the base of its tail. It grows to four inches.

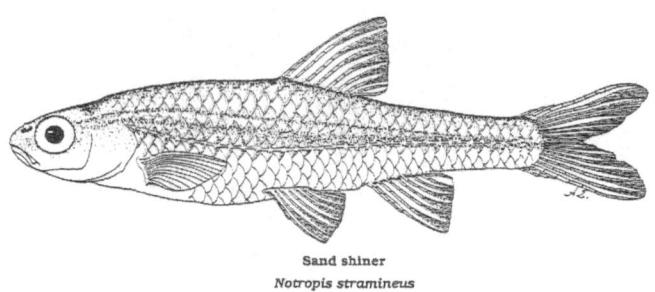

Sand shiner
Notropis stramineus

The Sand shiner is a nondescript cylindrical minnow. Its anus is not pigmented as it is in the mimic shiner. Its lateral band is poorly developed. It grows to less than three inches.

A Blacknose dace

11
Suckers – *Catostomidae* - sucking type ventral mouth, thick lower lip, toothless jaws, cycloid scales, head scaleless, soft rayed, abdominal pelvic fins. **Found:** White sucker, Quillback, Longnose sucker, Northern Hog sucker, Spotted sucker, Silver redhorse, Black redhorse, Golden redhorse, Shorthead redhorse.

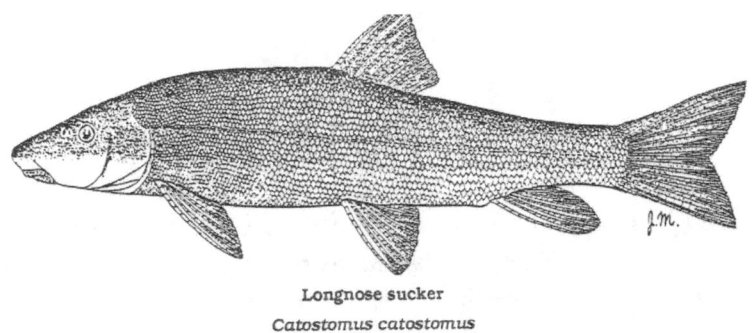

Longnose sucker
Catostomus catostomus

The Longnose sucker has a long and pointed snout extending far beyond the tip of its upper lip. It can grow to twenty inches.

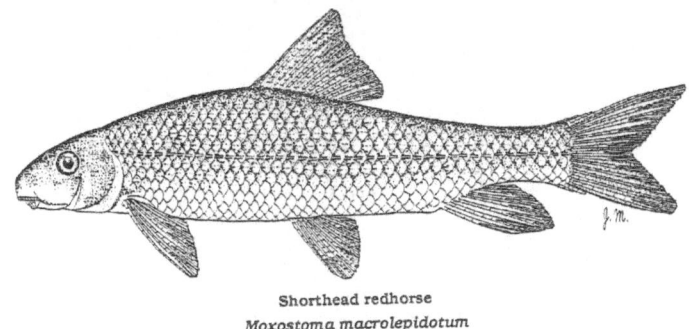

Shorthead redhorse
Moxostoma macrolepidotum

The Shorthead redhorse has a head convex between the eyes, its tail is reddish, its lips are striated, the lower lip might be broken into papillae, the head is small and short and its mouth is small. It is just short of twenty inches when mature.

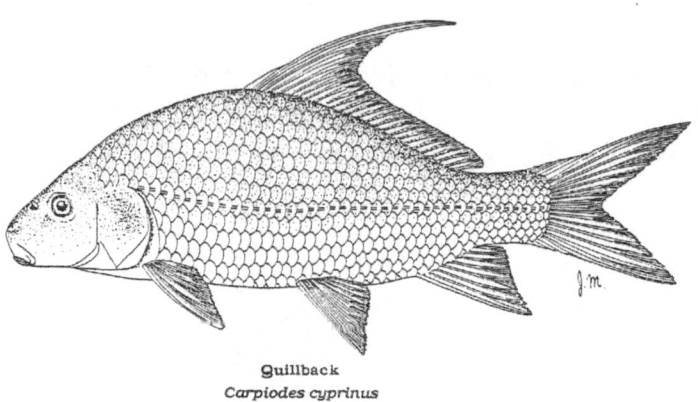

Quillback
Carpiodes cyprinus

The Quillback is easy to recognize with its moderately deep body which is not as deep as the carpsuckers. Its body is silvery and its snout is blunt. It grows to thirty inches.

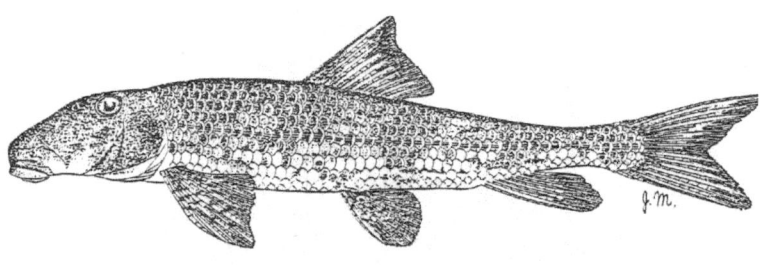

Northern hog sucker
Hypentelium nigricans

The Northern hog sucker has a cylindrical body that is not laterally compressed. It is deeply concave between the eyes. It has three oblique dark saddles across the back when it is young. Its sides have dark blotches and irregular spots. It grows to about ten inches.

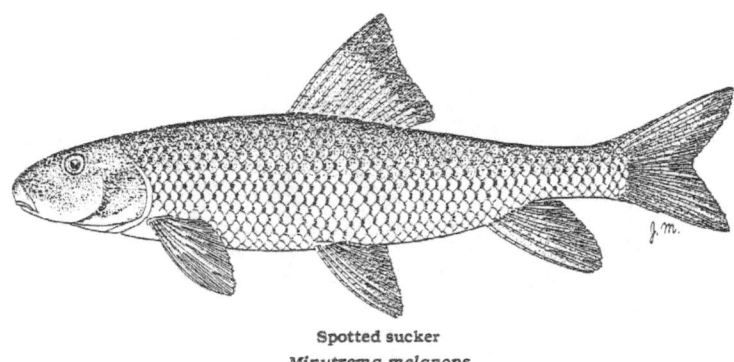

Spotted sucker
Minutrema melanops

The Spotted sucker has a black spot at the base of each side scale. Its body is slender and its mouth is horizontal. Its lips are thin and striated. It grows to fifteen inches.

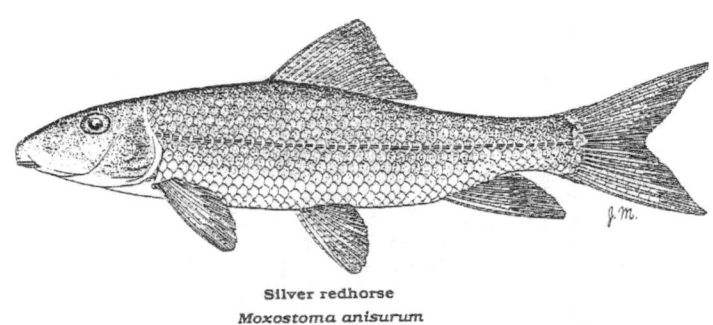

Silver redhorse
Moxostoma anisurum

The Silver redhorse is convex between the eyes. Its fold of lips are broken up by transverse creases. Its back has a moderately high arch to its dorsal fin. It grows to about two feet when fully mature.

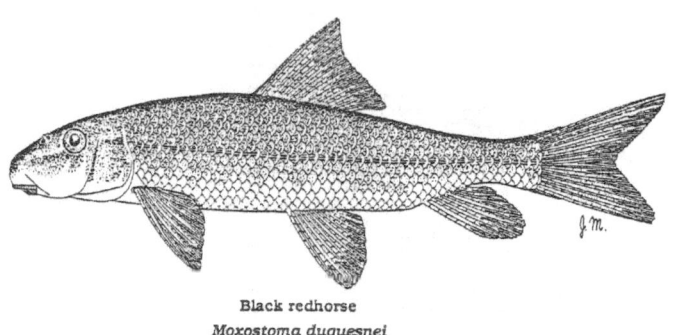

Black redhorse
Moxostoma duquesnei

The Black redhorse has a slightly concave dorsal fin. Its body is long and slender. Its upper lips are separated from the jaw by grooves. The head is convex between the eyes. The eyes are small. Its lips are striated. It grows to about sixteen inches.

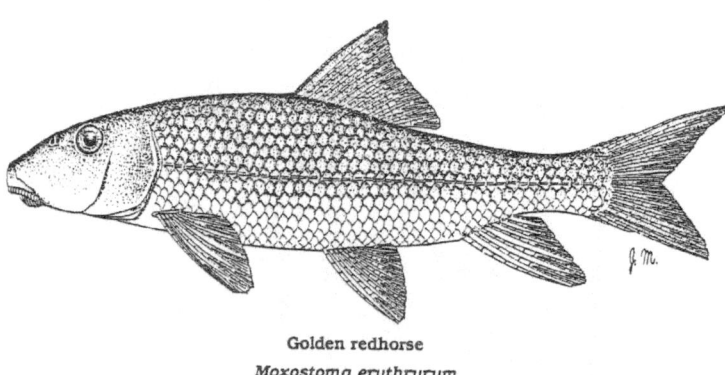

Golden redhorse
Moxostoma erythrurum

The Golden redhorse is convex between the eyes and its lips are striated. Its dorsal and caudal fins are light slate gray in color. It has little or no arch in its back. It grows to twenty inches.

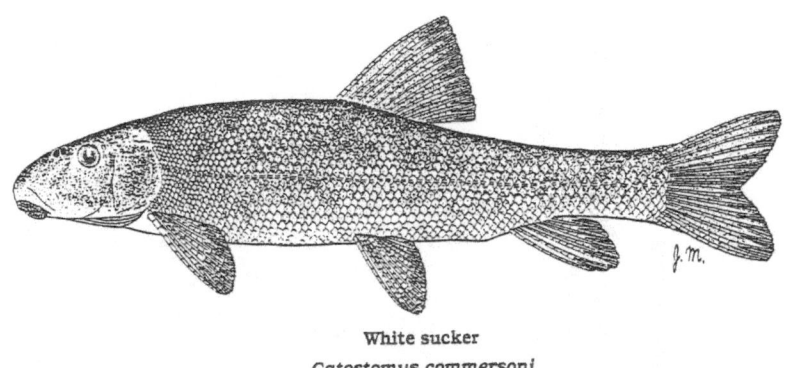

White sucker
Catostomus commersoni

The White sucker is convex between the eyes. Its scales are crowded toward the head. This is not characteristic of redhorses. There are 3 lateral V blotches on the young. The White sucker grows to about twenty inches.

12
Catfishes – *ictaluridae* – upper and lower jaws with barbels, chin with several barbels, scaleless, spine like hardening of fin rays at the front of pectoral and dorsal fins, soft rayed, adipose fin, single dorsal fin, pelvic fins abdominal. **Found**: Brown bullhead, Black bullhead, Yellow bullhead, Channel catfish, Mountain madtom, Stonecat, Tadpole madtom, Brindled madtom, Northern Madtom..

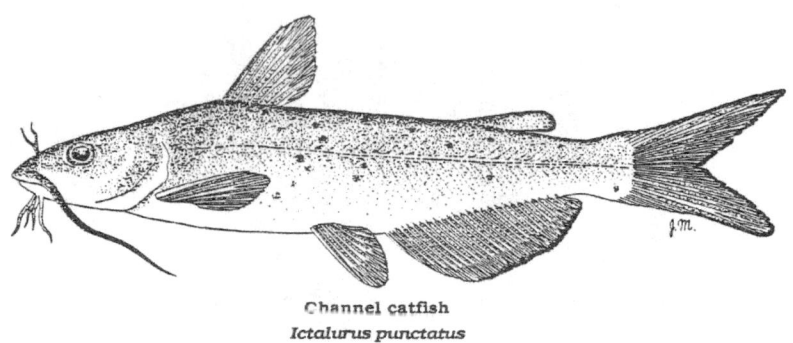

Channel catfish
Ictalurus punctatus

The Channel catfish has a deeply forked tail. Its upper jaw is longer than its lower jaw. It can grow to thirty inches.

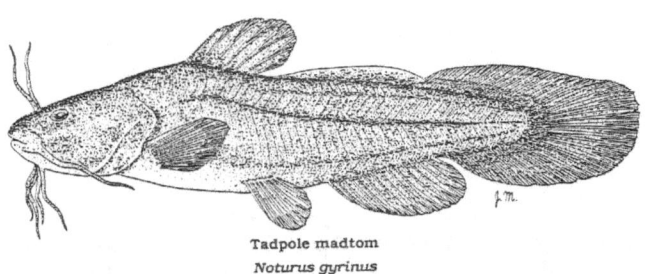

Tadpole madtom
Noturus gyrinus

The Tadpole madtom has a rounded tail. Its pectoral spines are not separated, dark streak along its lateral line, its a little thicker than other madtoms and it grows to a bit over three inches. The illustration is about as big as it gets. It can be mistaken for an immature version of a larger catfish.

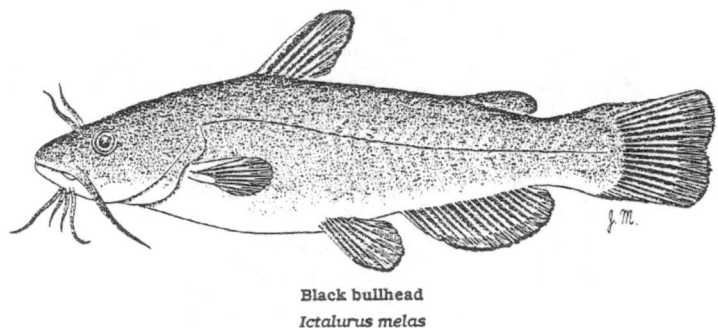

Black bullhead
Ictalurus melas

The Black bullhead has dusky to black chin barbels. Its pectoral spines are not serrated on the posterior edges. Its body is chubby and deep. It has black membranes on its pelvic and anal fins. It averages out to about twelve inches when mature although larger specimens have often been recorded.

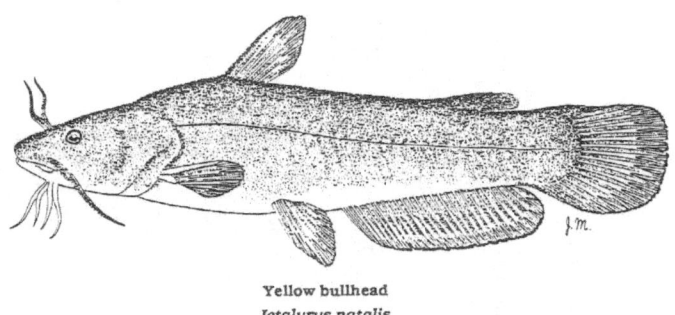

Yellow bullhead
Ictalurus natalis

The Yellow bullhead is of course yellowish in color. Its chin barbels are yellow or whitish. Its tail is rounded. It grows to about fifteen inches.

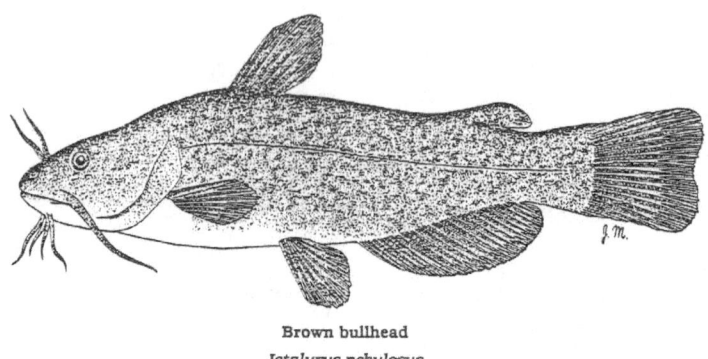

Brown bullhead
Ictalurus nebulosus

The Brown bullhead differs mostly in color from the yellow and the black bullheads. One could argue about it being a separate species. Its chin barbels are dusky. Its pectoral spines are serrated on the posterior edge. It grows to about twenty inches.

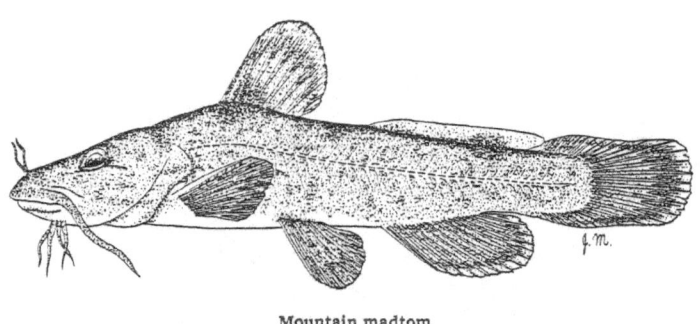

Mountain madtom
Noturus eleutherus

The Mountain madtom looks like a miniature catfish in its maturity. Its tail is squared. It has a low adipose fin attached to its back that is almost separated from the caudal fin. It has a dusky band on its dorsal fin. Its upper jaw is longer than its lower jaw. It grows to less than four inches.

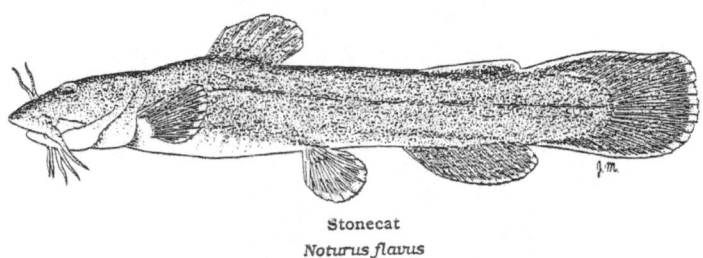

Stonecat
Noturus flavus

The Stonecat has a squarish tail with a light border. It has a long low adipose fin on its back and this is separated from the caudal fin by a notch. Its pectoral spines are not serrated. Its upper jaw is much longer than its lower jaw. It can grow to ten inches.

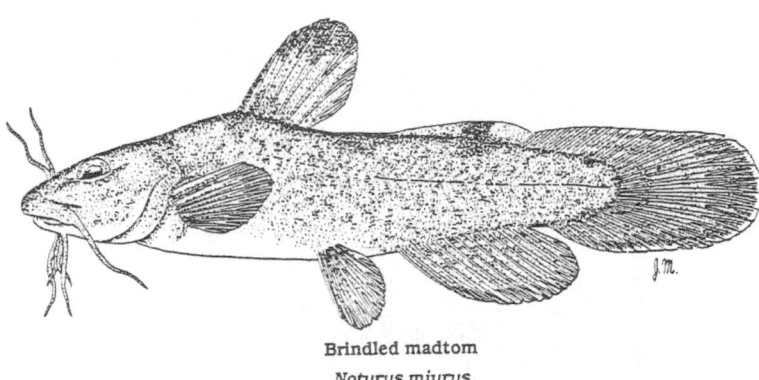

Brindled madtom
Noturus miurus

The Brindled madtom is a small catfish with a rounded tail. It usually has four saddle bands on its back. It has a large dark spot on its upper dorsal fin. Its upper jaw is longer than its lower jaw. It grows to less than four inches.

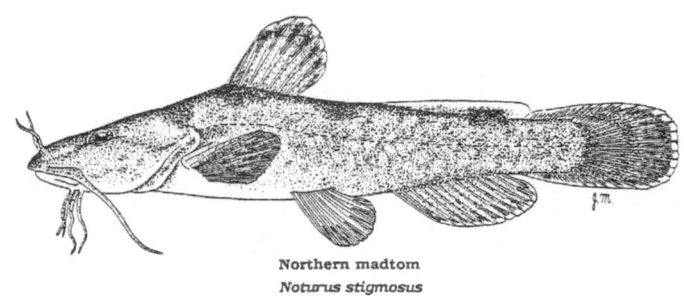
Northern madtom
Noturus stigmosus

The Northern madtoms tail is somewhat squarish. It has saddle bands on its back with a distinct one near the anterior base of the dorsal fin. Its caudal fin has a crescent-shaped band in the middle. Its upper jaw is slightly longer than its lower jaw. It grows to about four inches.

13
Trout-perches – *Percopsidae* – mouth small, jaws with teeth, ctenoid scales, head scaleless, spines small and inconspicuous, adipose fin, pelvic fins abdominal, upper half of the body is spotted. Found: Trout perch

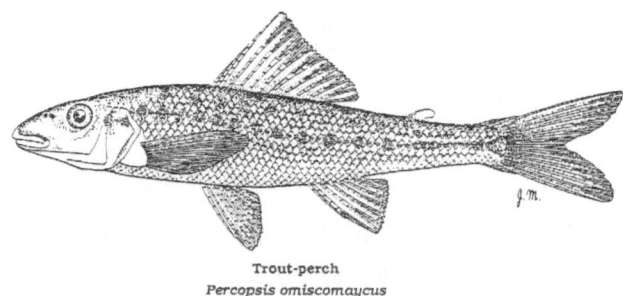
Trout-perch
Percopsis omiscomaycus

The Trout- Perch is not a trout or a perch. It's identified as a separate species. It is rare and grows up to seven inches.

14
Codfish – *Gadidae* – chin with single medial barbell near the lip, small cycloid scales which are embedded and do not overlap, soft rayed, long dorsal fin which may be divided into two or three parts, pelvic fins jugular, body not compressed laterally. **Found**: Burbot

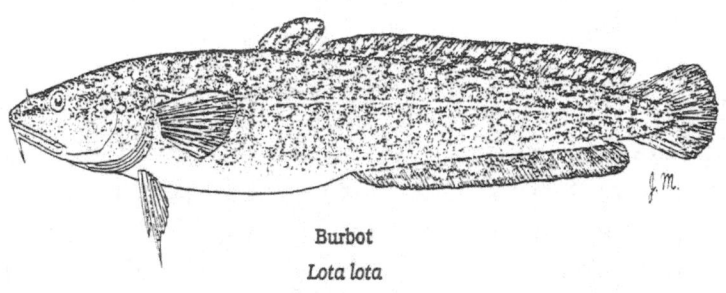

Burbot
Lota lota

The Burbot has two dorsal fins, the second one long, and a long anal fin. It is catfish-like in shape. It grows to thirty inches.

15
Killifish – *Cyprinodontidae* – head flattened on top, mouth opens upward, jaws short and with teeth, cycloid scales, head with scales, soft rayed, single dorsal fin, pelvic fins abdominal, tail region strongly compressed, tail rounded, lateral line incomplete or only partially developed, body with vertical bars or stripes in many species. **Example found**: Banded killifish.

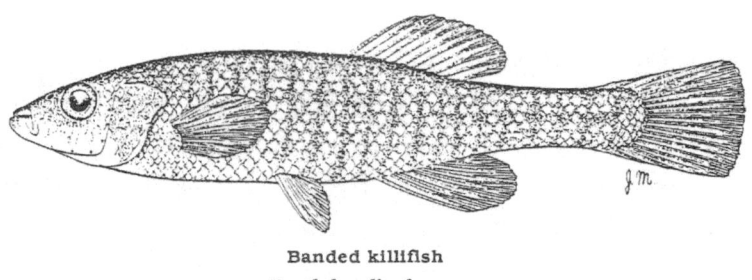

Banded killifish
Fundulus diaphanus

The Banded killifish has many lateral dark zebra-like bars. There is no lateral line. The adult only grows to four inches.

16
Silversides – *Atherinidae* – small, silvery, slender, somewhat transparent, conspicuous silvery band on its sides, dorsal fin divided into a small spinous portion and a larger soft rayed portion, anal fin is larger than either dorsal fin, mouth terminal and directed upward, large eyes. **Found**: Brook silverside.

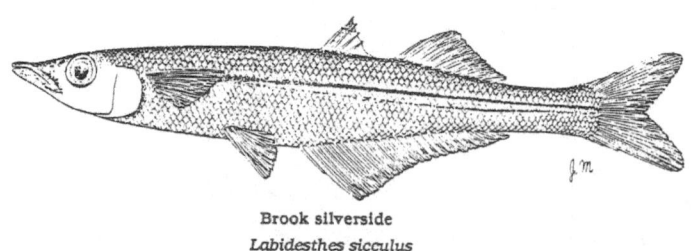

Brook silverside
Labidesthes sicculus

The Brook silverside is a small fish only growing to three inches, about actual size above. Its slender elongated body is an identifying factor as is its lower jaw slightly projecting past its upper jaw. Its caudal fin is distinctly forked.

17
Stickleback – *Gasterosteidae* - spiny rayed, isolated spines in dorsal fin, scaleless, pelvic fins thoracic and reduced to a single spine with no more than 2 soft ray rudiments, caudal peduncle slender. **Found**: Brook stickleback The Lake Superior list has several sticklebacks on it.

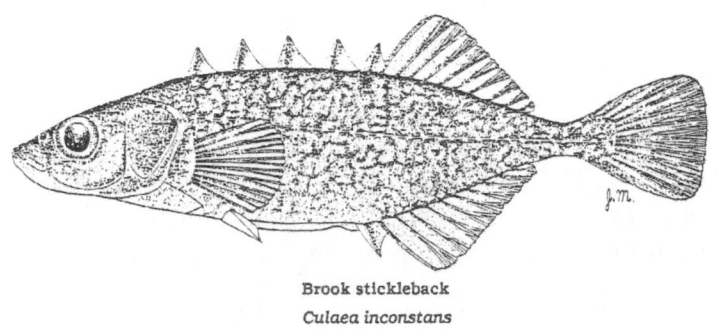

Brook stickleback
Culaea inconstans

Aptly named, the Brook stickleback has five spines, sometimes four, sometimes six on its back. Its caudal peduncle is deeper than it is wide. It grows to slightly less than three inches.

18
Temperate bass – *Percichthyidae* – jaws with teeth, opercles with a well developed spine, ctenoid scales, spiny rayed, dorsal fins separated or but slightly joined together, pelvic fins thoracic, sides with horizontal dark stripes. **Found:** White bass, White perch.

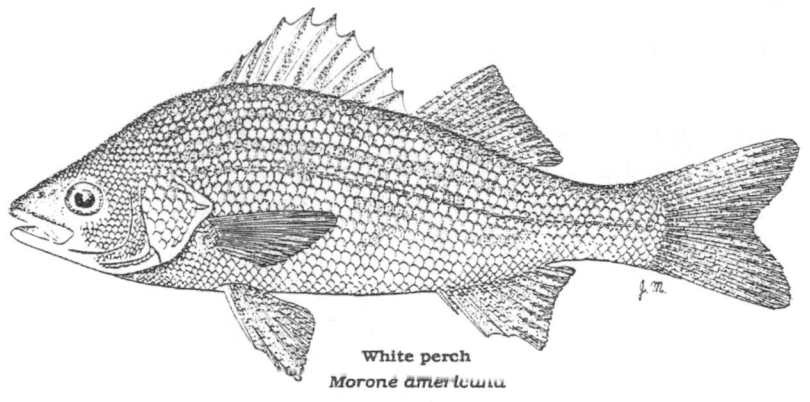

White perch
Morone americana

The White perch has jaws of equal lengths and dorsal fins slightly joined at the base. It has a rather small mouth and no distinct horizontal stripes. It can grow to twelve inches.

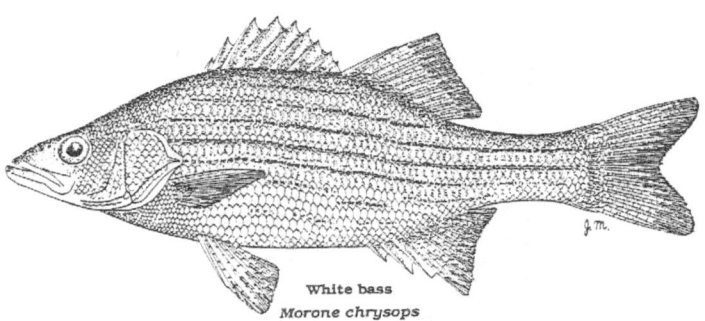

White bass
Morone chrysops

The White bass has separated dorsal fins. Its lower jaw projects in front of the upper jaw. It has about seven horizontal stripes on its side. These may be broken. The adult can grow to sixteen inches.

19

Sunfish – *Centrarchidae* – jaws with teeth, ctenoid scales, spiny rayed, dorsal fin single, dorsal fin almost separated in the largemouth bass, thoracic pelvic fins. **Found**: Bluegill, Rock bass, Green sunfish, Pumpkinseed, Warmouth, Longear sunfish, Smallmouth bass, Largemouth bass, White crappie, Black crappie.

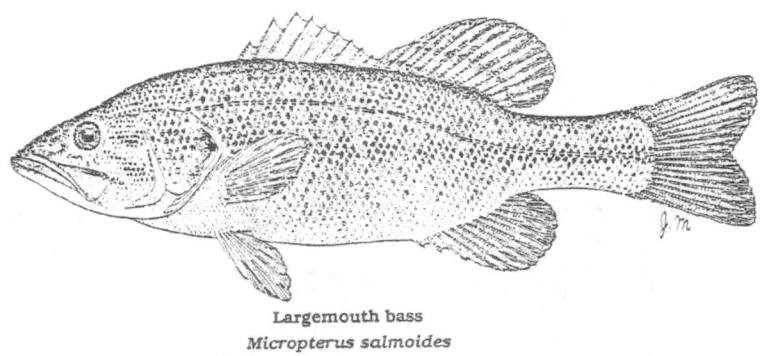

Largemouth bass
Micropterus salmoides

The Largemouth bass has an upper jaw that extends beyond the hind margin of the eye. The young have large eyes. The average adult is about twenty inches in length.

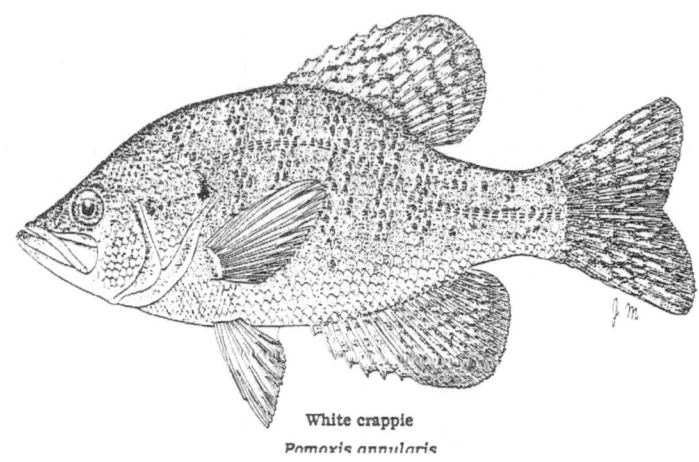

White crappie
Pomoxis annularis

The base of the White crappie anal fin is slightly larger than the base of the dorsal fin. Its sides has five to ten vague vertical bands. The adult length can reach sixteen inches.

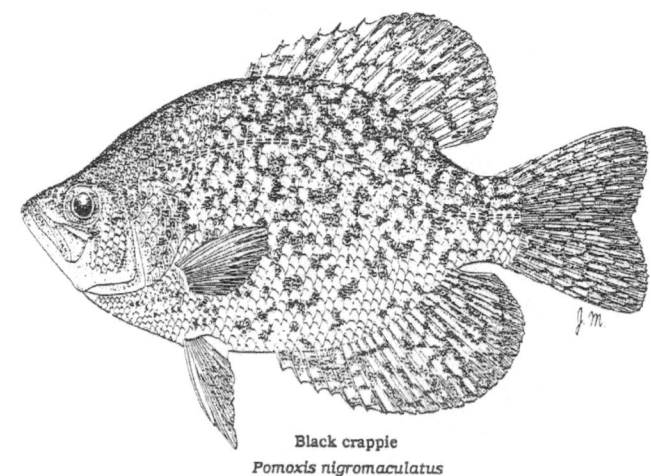

Black crappie
Pomoxis nigromaculatus

The Black crappie dorsal and anal fins are equal in length. It usually has six anal spines. It has a scattering of dark groupings on its sides. It can grow to about twelve inches.

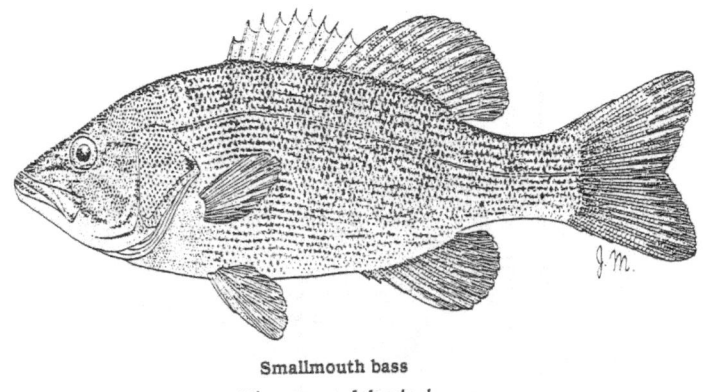

Smallmouth bass
Micropterus dolomieui

The Smallmouth bass gets as big as the largemouth bass. The smallmouth eyes extend beyond its jaws. It has a single dorsal fin with a shallow notch. The young have more dark markings than the adult. The smallmouth grows to less than twenty inches.

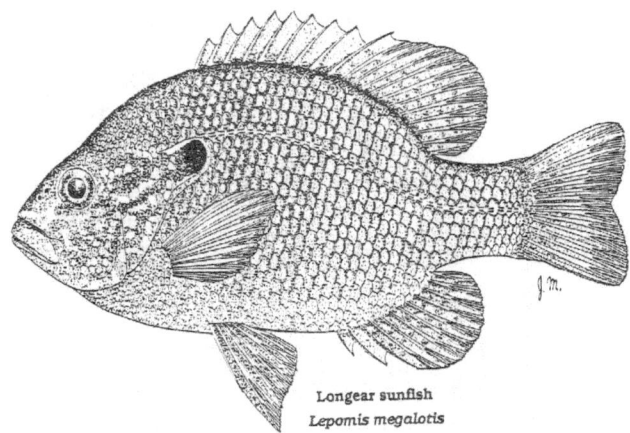

Longear sunfish
Lepomis megalotis

The Longear sunfish has an elongated opercle flap. The dark spot on the opercle has a red rim. It grows to eight inches.

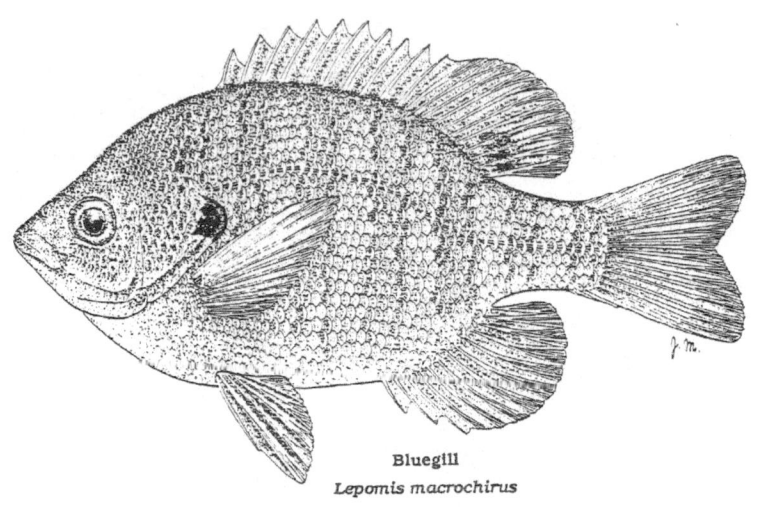

Bluegill
Lepomis macrochirus

The Bluegill has vertical bars on its sides and a dark blotch at the base of its dorsal fin posterior. Its mouth is small and its gill rakers are long. An adult may get to ten inches.

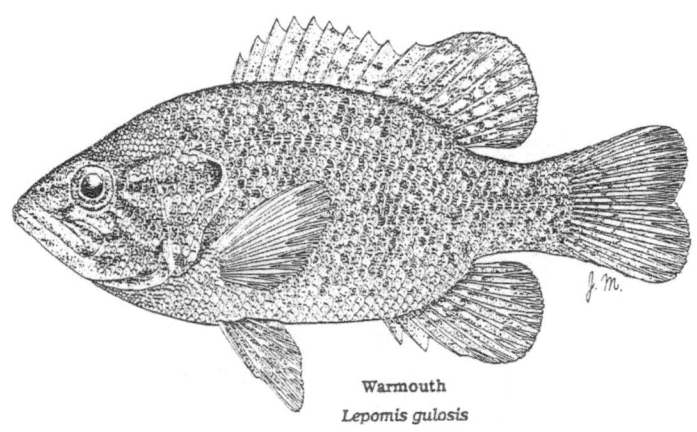

Warmouth
Lepomis gulosis

The opercle on the Warmouth has a dark lobe with a whitish margin. It has teeth on its tongue. The adult may get to ten inches.

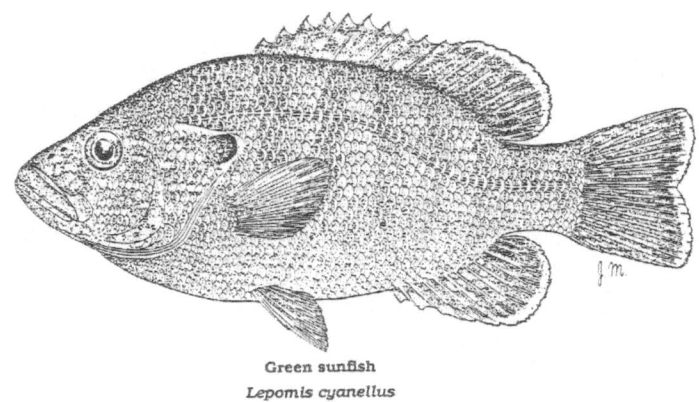

Green sunfish
Lepomis cyanellus

The Green sunfish has a shallower body and is thicker than the other sunfish. The head is about thirty percent of its length. It has three anal spines. Greenie grows to about six inches.

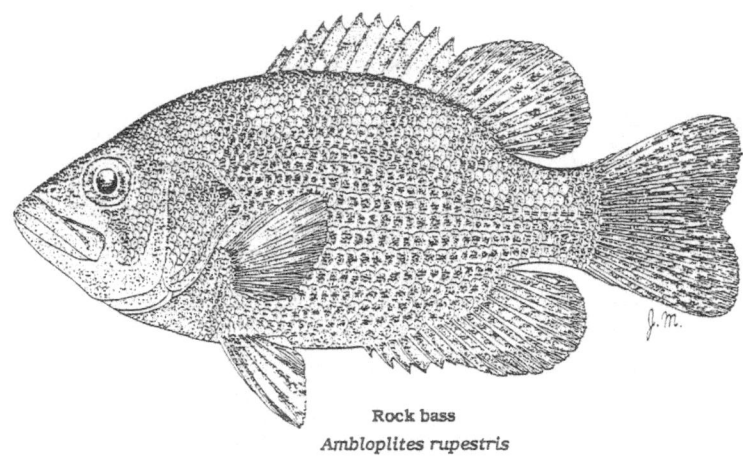

Rock bass
Ambloplites rupestris

The Rock bass usually has red eyes and brownish spotting overall. It grows to about twelve inches.

20

Perches – *Percidae* – jaws with teeth, clenoid scales, spiny rayed, two distinct dorsal fins, thoracic pelvic fins. **Found:** Yellow perch, Eastern sand darter, Greenside darter, Rainbow darter, Iowa darter, Fantail darter, Spotted darter, Johnny darter, Tippecanoe darter, Variegate darter, Banded darter, Logperch, Channel darter, Longhead darter, Blackside darter, Sauger, Walleye

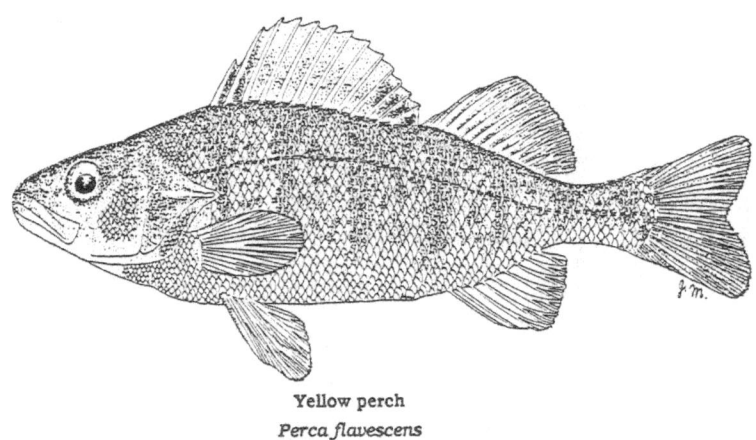

Yellow perch
Perca flavescens

The Yellow perch is one of the premier food fishes of the Great Lakes. Its body is yellowish with around seven vertical dark bars. It has a dark blotch on the webbing between the last four dorsal spines. It grows to around twelve inches.

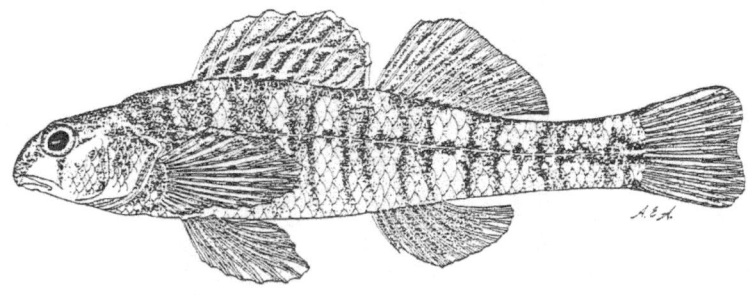

Banded darter
Etheostoma zonale

The illustration above is larger than any Banded darter. They have brownish spots along their lateral lines and eight narrow bands encircling their bellies. They grow to three inches.

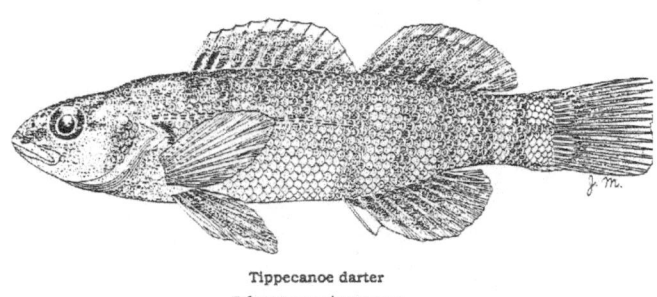

Tippecanoe darter
Etheostoma tippecanoe

The Tippecanoe darter has a pointed snout, not visible in the illustration. Its tail is usually slightly notched. Its mouth is slightly oblique and its cheeks are almost scaleless. It only grows to about two inches.

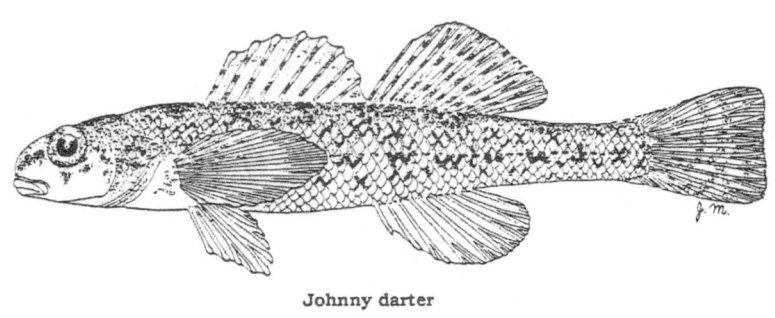

Johnny darter
Etheostoma nigrum

The Johnny darter has dark "W" shaped lateral markings. Its nape, cheeks, and breast are scaleless and its lateral line is nearly complete. It grows to less than three inches.

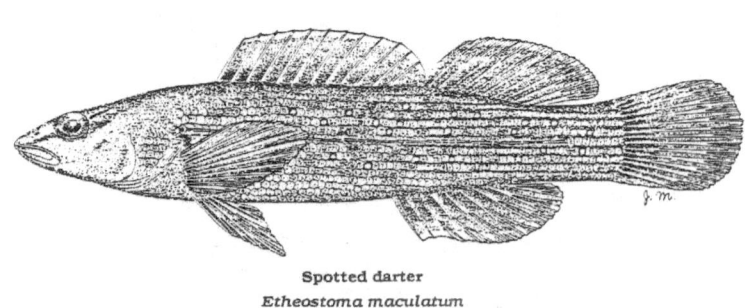

Spotted darter
Etheostoma maculatum

The male of the Spotted darter has red spots on its body. Its snout can be described as sharp. Its upper lip is connected to its jaw by a central bridge of flesh. It grows to slightly over three inches.

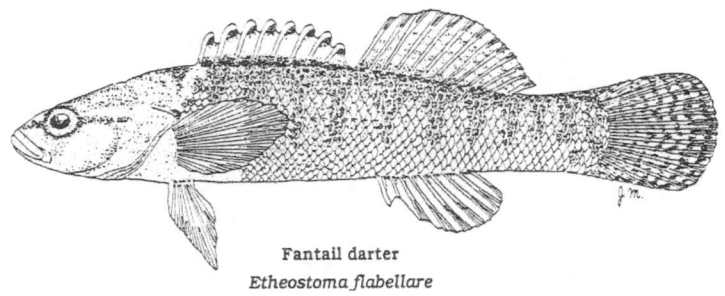

Fantail darter
Etheostoma flabellare

The Fantail darter has around ten short dark vertical bars on its sides. These are often vague, but they are there. The snout is straight and not rounded. The dorsal spines of the male have soft knobs. It grows to about three inches when mature.

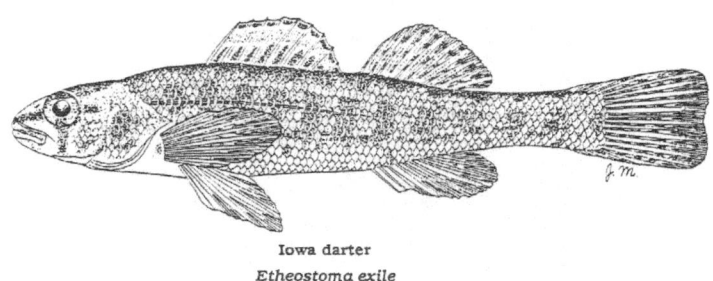

Iowa darter
Etheostoma exile

The Iowa darter has nine to eleven blotches on its side which are a reddish color in the male. Its cheeks are fully scaled with a teardrop below its eye. It grows to less than three inches.

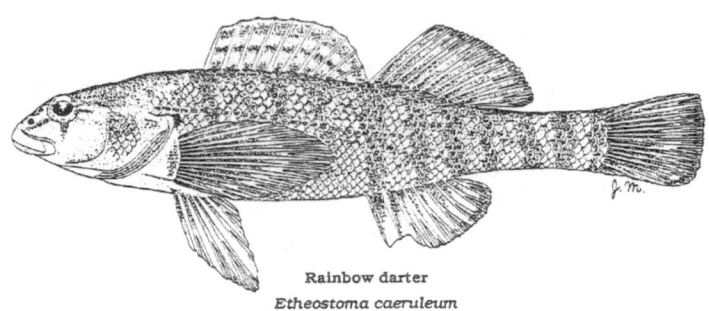

Rainbow darter
Etheostoma caeruleum

The Rainbow darter has around ten dark bars on each of its sides. Its snout is somewhat pointed and projects beyond its lower jaw. The adult length seldom reaches three inches.

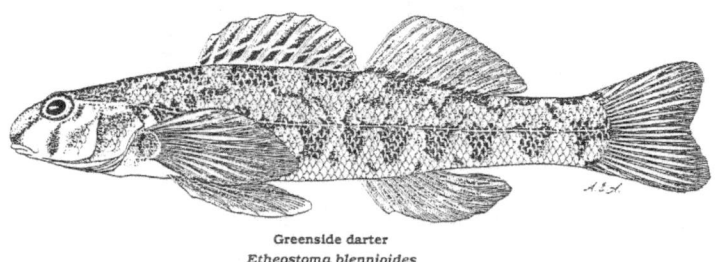

Greenside darter
Etheostoma blennioides

The Greenside darter has five to seven V shaped blotches on each side. Its snout is broadly rounded and seems to bulge. Its mouth is small. Its lateral line is straight. It grows to about four inches.

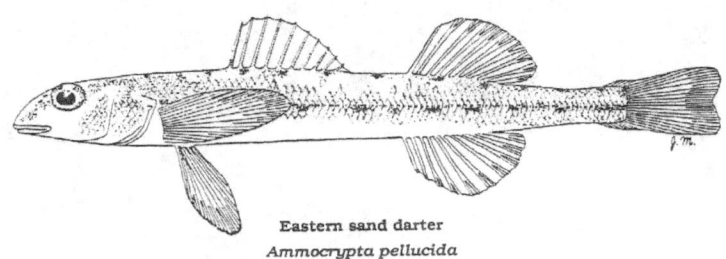

Eastern sand darter
Ammocrypta pellucida

The Eastern sand darter has about a dozen small round spots along the mid-line on each side and more than a dozen along the mid-line on its back. Its lateral line is complete and its snout is downward curved. It looks thin. It grows to less than three inches.

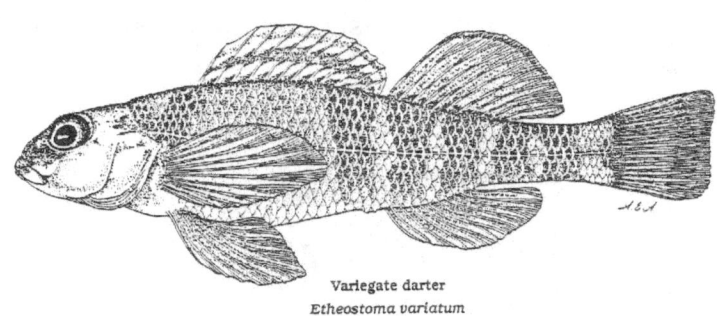

Variegate darter
Etheostoma variatum

The Variegate darter has a half dozen cross bars posterior located on its sides starting at the posterior end of the first dorsal fin. Its lateral line is complete and nearly straight. It grows to slightly less than four inches.

The head of a Logperch fish. Compare it with the sketch below.

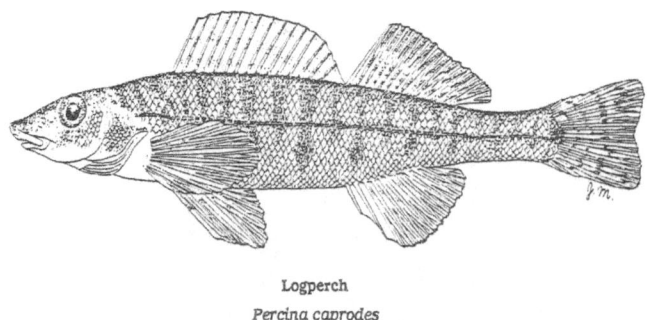

Logperch
Percina caprodes

The Logperch has a snout that protrudes beyond its upper lip. It has many irregular vertical bars on its sides. Its maximum length is about six inches.

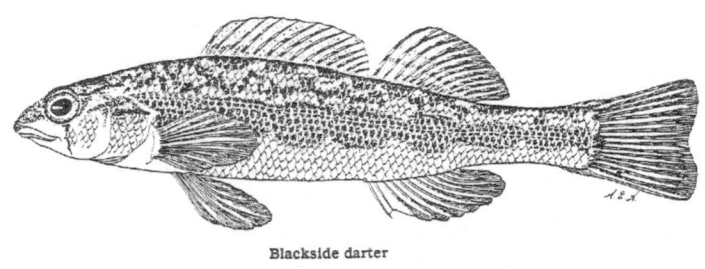

Blackside darter
Percina maculata

The Backside darter has a half dozen elongated blotches on its sides. It grows to about four inches.

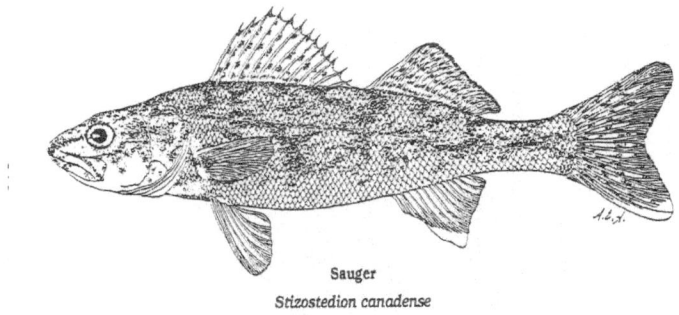

Sauger
Stizostedion canadense

The Sauger has a forked tail and separated dorsal fins. It is often referred to as a small walleye. It has large sharp almost canine teeth and round dusky spots on its spiny dorsal fin. It grows to less than twenty inches. It can tolerate less than pure water better than most other fish species closely related to it.

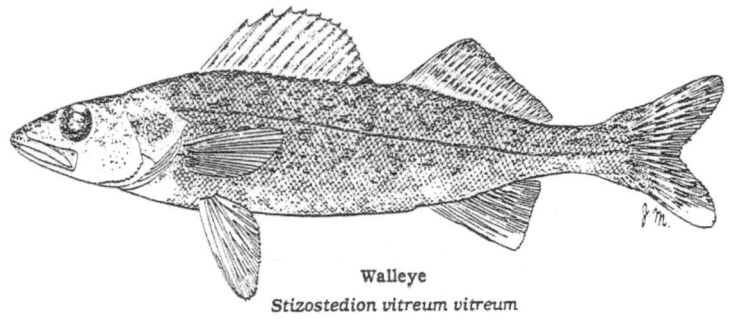

Walleye
Stizostedion vitreum vitreum

The Walleye competes with perch for being the most desired pan fish. Its body is brassy and mottled, never bluish which presents an argument about blue pike being a separate species. It has large eyes high on the sides of its head. The lower lobe of its caudal fin is whitish. It grows to thirty inches. It has several names such as mooneye and ghosteye.

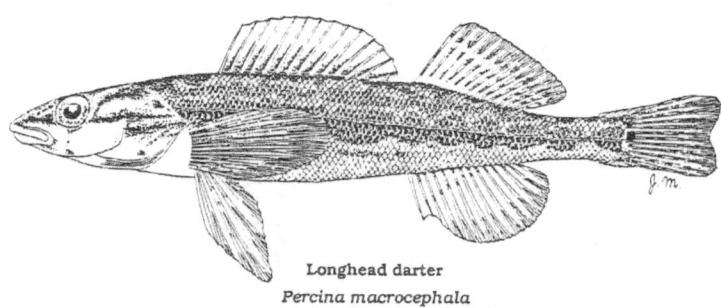

Longhead darter
Percina macrocephala

The Longhead darter has blotches on its side which merge into a continuous band. Its snout is long and its body is slender. Its upper lip is connected to the jaw by a central bridge of flesh. It grows to about four inches.

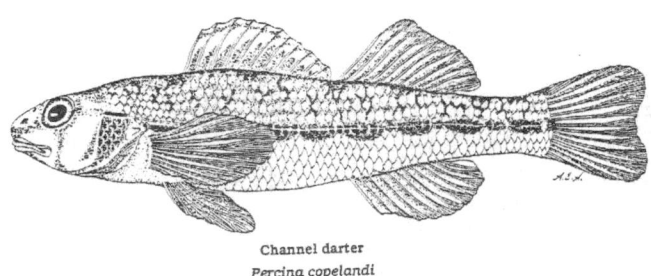

Channel darter
Percina copelandi

The Channel darter has a row of small blotches on its sides. Its snout is blunt. It grows to almost three inches.

21
Drum – *Sciaenidae* – lateral line extends across the caudal fin, teeth are molar like, ctenoid scales, spiny rayed, dorsal fin long and continuous, pelvic fins are thoracic. **Found**: Freshwater drum.

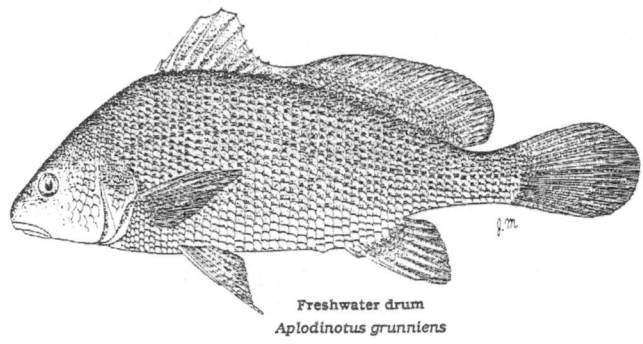

Freshwater drum
Aplodinotus grunniens

The Freshwater drum has a blunt snout and a subterminal mouth. Its body is deep and slab-sided. It grows to thirty inches. It is also known as sheepshead.

Sculpins – *Cottidae* – very large flattened head, scaleless with prickles only spiny rayed, pre-opercular spine, large pectoral fins, thoracic pelvic fins. **Found**: Mottled sculpin

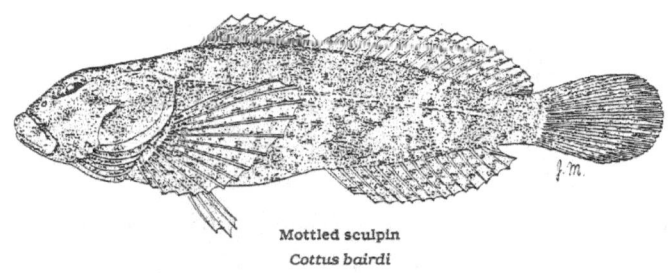

Mottled sculpin
Cottus bairdi

The Mottled sculpin has a dark and mottled body and 4 pelvic ray fins. Its dorsal fin has a dark spot posteriorly. Its lateral line is incomplete. It grows to about four inches in length.

The Lake Superior listings of 2017 (found at the end of this work)

Generally, the major species found in Lake Erie in 1992 were also found in Lake Superior. These are listed near the back of this work according to common names with the scientific name given for each listing.

Listed below are the common name species not found above in the Thomas study. However, what Thomas calls Lake Herring, the Lake Superior study calls Cisco. So we must be wary of common names when we discuss species.

Atlantic salmon   Bloater   Cisco   Eurasian rufe
Kivi   Fourspine stickleback, Ninespine stickleback
Splake. Threespine stickleback Round Goby   Tubenose
Goby Shortjaw Cisco

The Bloater is a type of herring. It is usually described as "rare."

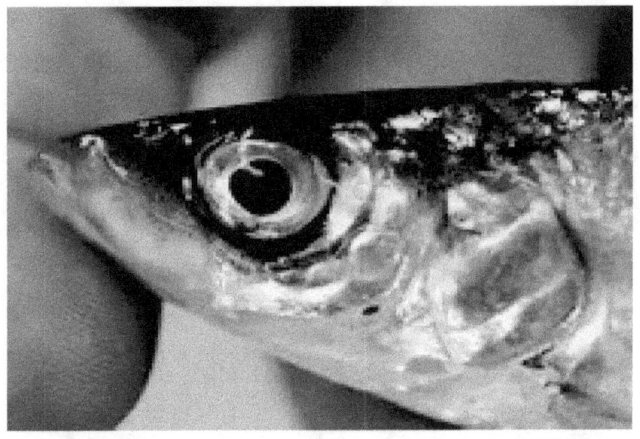

Head of a Bloater. Notice the herring eye and snout.

Herring type fish in the Great Lakes are Bloater (*Coregonus hoyi*), Cisco (*C. artedi*), Kivi (*C. kiyi*), Lake Whitefish (*C.clupeaformis*) and Shortjaw Cisco (*C. zenithicus*).

The splake is a hybrid fish created by the union of lake trout and brook trout. **Note:** When I was fishing in one of the lakes in the Haliburton Highlands of Canada a low flying airplane spilled out a bucket of fish. I was told they were stocking the lake with the newly developed Splake. Many years later, I went back to those lakes and indeed the splake had taken hold.

A 24 inch splake

Eurasian ruffe

The Eurasian ruffe (Gymnocephalus) is an invasive species that has been considered a nuisance in the Great Lakes. However, one person's nuisance might be another person's pleasure.

The cisco can reach thirty inches in length and thrives in cold, clean water. (Coregonus) The three types found in Lake Superior are Common cisco, Shortjaw cisco, and Kivi.

The stickleback is an interesting little fish. The Brook stickleback was found in Lake Erie. Lake Superior has the

Threespine Stickleback, Fourspine stickleback and the Ninespine stickleback, sometimes referred to as the Tenspine stickleback. I guess, it depends on who is counting.

There were two species of goby found in the upper lakes, the Round Goby and the Tubenose Goby. The Goby was not in the Great Lakes when Professor Thomas completed his study in 1992. The first Goby was discovered in Lake Huron where it was believed to have been delivered in ballast water coming from a ship registered in the Ukraine. An early Goby was also discovered in Lake Erie.

NON NATIVE SPECIES – invaders?

The following fish are considered to be invasive species in the Great Lakes. This list also includes hybrid fish and introduced species. - American eel, American brook lamprey, Brook silverside, brown trout, Chinook salmon, Coho salmon, Common carp, fourspine stickleback, Freshwater drum, goldfish, Pink salmon, rainbow smelt, rainbow trout, round goby, sea lamprey, splake, threespine stickleback, tubenose goby, white perch.

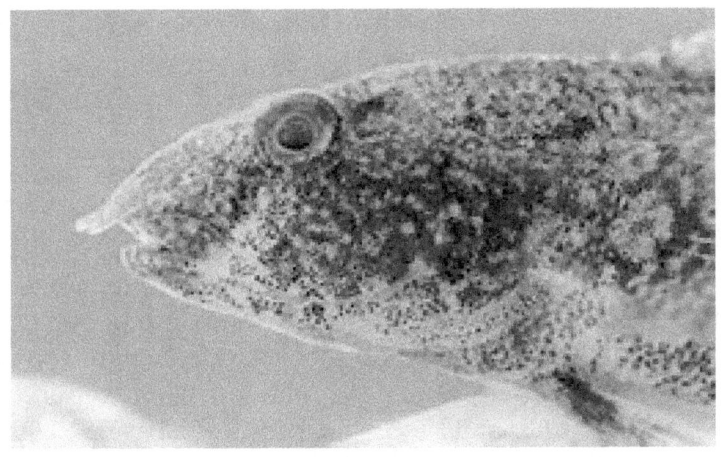

The tubenose goby

III. Practical Classifications

I had wrestled with my own system of classifying fish which was mainly concerned with utility rather than science. So in the final analysis I came up with fish species dividend into Sport Fish, Pan and Food Fish, Little Fish, Lamprey, and Odd Fish that didn't seem to fit any of the other groups.

1. Sport Fish

Sport fishing belongs in the same recreational realm as camping, biking, bird watching and outdoor photography Most anglers who indulge in the activity do not keep the fish they catch but phonograph it and quickly return it to the water.
    One may arrange the sport fish into sub-categories. I have three, **1.** sport only **2.** sport and food, and **3.** sport contest.

In the first category I place Muskellunge, Northern Pike, Grass Pickerel, Chain Pickerel and Tiger Muskellunge. The Tiger Muskellunge is a hybrid created by genetic manipulation and egg experimentation.

In the second category I place the bass, salmon, perch and trout. These are the target of the majority of people who fish for food as well as sport.

The third category is an abomination to the relationship between humans and nature and takes the form of contests and tournaments, usually based on species, size, weight and number of fish caught. I participated in one tournament and observed two others. My partner and I took the three large bass we caught home but many of the other entrants left theirs to rot. I suppose these fed gulls and raccoon while stinking up the area.

In one of the other tournaments the guts, heads and tails of many fish were left on shore after the participants cleaned their catch and departed. First class tournaments do not let this happen.

Upside: Tournaments do identify the anglers who have special skills in that activity. These winners help to encourage the sport of fishing and they also help to sell specific products as well as develop new products.

Muskellunge coloration can vary slightly, sometimes causing confusion as to its species. The average catch size is 45 inches.

The Northern Pike is often found in clusters. If you find one, you are apt to find others in the area.

Grass Pickerel

Chain pickerel

Some notes on Chain Pickerel

The Chain pickerel is also known as the "gunfish", "federation pike" or "federation pickerel". Pickerel is often a name given to walleye, although the true name belongs to the chain pickerel. Common nicknames are the "southern pike", "grass pike", "jack", "jack fish", and "eastern pickerel". The nickname "grass pike" is acceptable in general terms but not in science where local names are usually not recognized.

The chain pickerel has a distinctive, dark, chain-like pattern on its greenish sides. Its body outline resembles that of the northern pike. It may reach up to 30 inches long only on rare occasions. The gill area and cheeks of the fish are entirely scaled. The average size for chain pickerel is 24 inches with a weight of 3 pounds. Studies indicate the average chain pickerel caught by fishermen is under 2 pounds. Studies also show it has a life expectency of around 8 years.

Chain pickerel are rare in the Upper Great Lakes. That is according to researchers. I can attest to the fact that they are also rare in the Lower Great Lakes. The chain pickerel feeds primarily on smaller fish, like the northern pike, until it grows large enough to ambush large fish from cover with a rapid lunge and to secure it with its sharp teeth. Chain pickerel are also known to eat frogs, worms, water bugs, crayfish, and a wide variety of other foods. It is not unusual for pickerel to leap out of the water at flying insects, or even at dangling fishing lures. I once witnessed a Chain pike caught on a lure made up of a live mouse in a harness. It was an experiment.

The chain pickerel is a popular sport fish. It is an energetic fighter when hooked. Anglers, not me, have success with live minnows, spinner baits, spoons, plugs, and flies. I have seen them caught on bucktails and once with a feather embedded in a wooden spoon. If the angler intends to release a fish, it is advisable to use pliers to flatten the barbs on the lure's hooks before casting. Chain pickerel can swallow an entire lure, so it will be much easier to free a deeply hooked fish and get it back into the water as soon as possible. I hope no one intends to keep a Chain Pike, although most of them make a very good mount on a wall.

Practically any bass lure can be effective for pickerel, although like most pike, they seem to be particularly susceptible to flashy lures which imitate small forage fish. Dragging a plastic worm, lizard, frog, or other soft imitation can also be extremely effective.

A steel leader is necessary for sharp-toothed and active fish at two to three pounds. The angler would also do well to use 12- to 17-lb-test line on an open-faced spinning reel. Methods are similar to those for bass, such as dragging a lure through weeds in shallow water and jerking it side-to-side to give it the look of injured prey. Chain pickerel are voracious and opportunistic feeders, and will attack most any moving item that passes into their range of vision. The trick is to find out where they live.

Chain pickerel can grow to thirty inches and this one seems to have maxed out.

A selection from the field notes of Professor Thomas where he explains his reasons for some of his notes.

"The field characteristics given for the fishes in this guide are, for the most part, those that can be used in the field without dissection or the aid of a microscope. Body color has been used very little because it varies among different habitats and is usually destroyed in preservatives. Ray counts, with few exceptions have not been used as ray counting is a tedious job in the field, especially on small fishes. The adult length range for a species does not include the smallest or largest specimens ever collected; the inclusion of such individuals would make the length range character of little use in field identification. Thus, the length range is somewhat conservative but should include most adult individuals encountered."

2. Pan and Food Fish

This category of fish is for fun and profit. These are the good eating fish and are easily caught if one has patience and on a day when they are biting. Ha. Ha. These are the perch and sunfish near the top of the water and the catfish and suckers at the bottom They can be taken with lures, but the use of minnows and worms have been historically the most productive for the top feeders and worms for those at the bottom.

Perch

The perch grouping of fish includes what seems to be a wide variety but their characteristics of jaws with teeth, type of scales, spine rays and two distinct dorsal fins indicate their close relationships.

The large fish of this group include the Yellow Perch with its beautiful yellow and black body, the Logperch with its protruding snout, the Sauger with its forked tail and its bigger image the Walleye which does not have the distinct cross bars found in Yellow Perch.

A good ten inch Yellow perch.

Blackside darter. A small member of the perch group of fish. It grows to four inches at most. Darters are small fish with the characteristic perch features.

The sauger is a typical perch-type fish. Its eyes resemble the walleye and it is sometimes confused with that species.

Walleye may grow to 30 inches in length. The walleye is a good example of confusing common names without the use of scientific names. It is known by several regional names such as mooneye, ghosteye and spook fish.

A twenty inch Walleye and the lure that got it.

A sixteen inch sauger. Notice the fins and the ghost eye.

A Trout-Perch is not a trout and not a perch but it does have similar external physical characteristics.

Notes on Walleye

Few fish match the walleye for the combination of recreational angling enjoyment and table fare. Walleye can be found in the Great Lakes and connecting waters as well as inland rivers, lakes and reservoirs, though many of the inland fisheries are dependent upon stocking programs to maintain legal fishing populations. Walleye can be caught in numerous ways. Early in the season, fishing bottom with lead-head jigs tipped with minnows or with plastic grub bodies is the top technique, but as the season progresses, trolling with plugs or spoons or with spinners and crawler harnesses becomes the preferred method. Slow trolling baits at a variety of depths can be productive although walleye are usually associated with the bottom.

The most active fish are sometimes suspended in the water. The walleye can readily be taken on live bait; nightcrawlers drifted along the bottom, leeches suspended under a slip bobber or minnows fished on a tight line will all produce. In fall, jigging with spoons in deep water is a popular technique. Walleye can be photosensitive. Fishing is often best early and late in shallow water, though that is less critical in deep water. But walleye often move shallow to feed at night and casting with artificial lures or drifting with live bait will all produce walleye after dark.

Walleye are popular quarry for ice fishermen who jig with artificial baits such as lures or spoons, often tipped with minnows on slip-bobber rigs or with tip-ups baited with live minnows. Walleye fishing through the ice usually begins and ends in shallow water areas with deep water more productive during the heart of the winter. *Adapted from an article of the Michigan Department of Natural Resources.*

Research Note: A fish does not add new scales as it grows, but the scales it has increase in size. In this way, growth rings are formed and the rings do reveal the age of the fish.

Trouts and Salmon

Trout and salmon are distinguished by a small fin located between the dorsal fin and the tail. They also have heads without scales. They have low abdominal fins in the middle of their bodies. These make up some of the largest fish in the lakes.

The trout and salmon group living in the lakes consist of Lake Herring, Lake Whitefish, Coho Salmon, Rainbow Trout, Sockeye Salmon, Chinook Salmon, Brown Trout, Brook Trout and Lake Trout.

The Lake Herring is silver in color and has an almost toothless jaw. Lake Whitefish have a blunt snout and toothless jaw. Coho Salmon are easily recognized by the small black dots above its lateral line. Rainbow Trout, of course, has beautiful rainbow hues. It does not have reddish spots. The Steelhead is a rainbow trout that lives out in the lake and comes into tributary streams to spawn. An old saying goes, "The Steelhead is a rainbow trout that has gone to sea."

Sockeye Salmon do not have black spots on its caudal fin. It is bluish above and silver below. Chinook Salmon have a sharp snout with teeth. All of its fins have some black spots. Brown Trout have lateral red brown spots which appear speckled. The Brook Trout is one of the smallest of the group and its back markings are often described as worm-like. It is rare in the lakes.

Lake Trout can mature to three feet in length. They have a lateral line and coloration resembling brook trout but without the red-brown color. Lake Trout are the primary prey of the Sea lamprey.

A mess of lake herring.

A mess of lake whitefish

Coho salmon adults are 18 to 24 inches long. They were identified in the Great Lakes over a hundred years ago. Today, they are stocked.

Sockeye salmon reach 24 inches in length. There is a smaller version, known as Kokanee which is not in the lakes.

Chinook salmon were introduced to the Great Lakes in 1873, When I began fishing in the lakes in 1951 I never heard of anyone catching a Chinook salmon. The states bordering the lake began stocking the salmon in 1966 and they took hold. There are some questionable reports that they have, on some occasions, spawned in tributary streams of the lake. They, like their cousin salmons, are a stocked fish.

Some notes on salmon stocking

Salmon are spawned at hatcheries during the month of October. The eggs hatch out in late November through December. Most salmon are stocked as 3 inch fingerlings in May or June. Besides shore stocking, some of the salmon are also pen stocked.

Pen stocking is a cooperative effort between the state fish departments and area sportsman groups. Pen stocking allows recently stocked fish a chance to acclimate to their new surroundings and offers some protection from predators. Fish and feed are provided by the state fisheries while the sportsmen's groups build and maintain the pens and feed and take care of the fish for approximately 3 weeks.

The fish are subsequently released into the stream or bays. This pen stocking program has been very successful. Shortly after stocking, salmon "smolt" and imprint on the scent of the stream before migrating downstream to the lake. Coho are stocked as either pre-smolt fall fingerlings at 10 months of age (4½ inches long) or as 6 inch yearlings at 16 months of age. The life history of the coho salmon requires that they stay in the streams for at least one year before smolting and moving down to the lake.

Once they reach the lake, salmon grow rapidly on a diet of alewives. Chinook salmon returning to the rivers where they were stocked range in age from 1 to 4 years. Age 2 and 3 fish make up 90% of the run and will weigh between 15-25 pounds. Mature coho salmon return to spawn as age 2 and will average 8-10 pounds.

Maturing salmon begin to "stage" off the tributary stream mouths from mid-to-late August. By early September some fish have usually started to trickle into the tributaries. The peak of the run when the best stream fishing occurs is actually a rather short 4 week period

On some salmon streams entering the lake the timing of the runs is more dependent on rainfall. Generally salmon will enter these streams somewhat later with the peak occurring in mid-October.

Once salmon enter the streams, they are no longer feeding. Their bodies are undergoing rapid physiological changes and their sole purpose left in life is to spawn. While they are not actively feeding, they do exhibit several activities which make them vulnerable to traditional sport fishing techniques. One of these activities is aggression or territoriality, and another is their attraction to fish eggs or egg shaped lures.

Although salmon stocking has been going on for a number of years the programs are still, more or less, in an experimental stage. There is much more to learn about the raising and transfer of ocean fish to the great lakes. As long as people purchase fishing licenses, the program will continue.

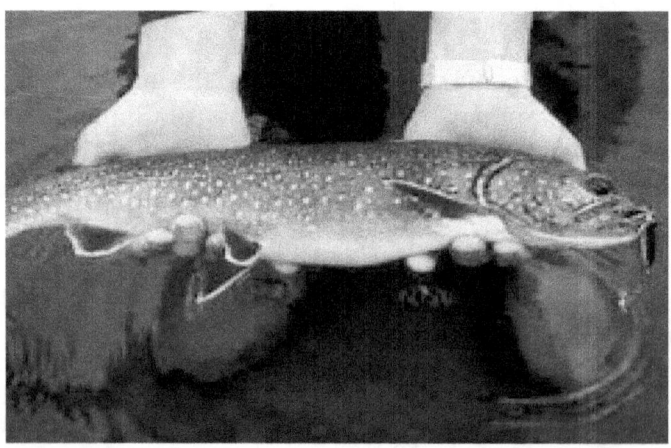

A splendid lake trout. The Lake Trout lives in the deepest water of the lake. It can grow to 36 inches.

Brown trout are about eight inches long in tributary streams but grow to 24 inches out in the lake. Where do you think this one was caught?

The beauty queen, Rainbow Trout, Steelhead, is around eight inches long in streams but 20 inches long in the lakes.

Fishing for Lake Trout

Lake trout (Salyelinus namaycush) is a trout that is native to North America. They live in freshwater and are one of the most popular game fish in North America. More than 25% of the world's lake trout population resides in Lake Ontario. Lake Erie also has had a good survival rate of lake trout. Their diet varies based on the length, weight and age of the fish, but includes small crustaceans, snails, leeches, insect larvae and other small fish. Due to popularity among anglers lake trout populations can be decimated and they need to be supported by stocking programs.

Lake trout prefer to live in large lakes that are deep with cold water. Generally they spawn during the fall season, but this can be affected by the location of the lake and weather patterns. Juvenile trout are reclusive and will seek out deep water where they feed on plankton and other small aquatic prey. Lake trout are easy to distinguish from other trout by their color variation, which is usually yellowish to white spots on their dark green to gray body. They have an elongated, stout head like all other trout. Their belly is white and at times can have an orange-reddish color paired fins.

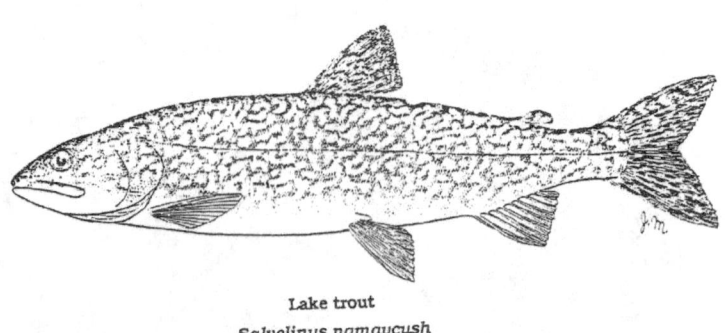

Lake trout
Salvelinus namaycush

Lake trout and Brook trout are similar in appearance. The Lake trout does not have the red color and lower fins without black stripes. The Lake trout body is covered with light spots on a dark background. Its caudal fin is deeply forked, more so than on the Brook trout. The Lake trout can grow to thirty six inches or more.

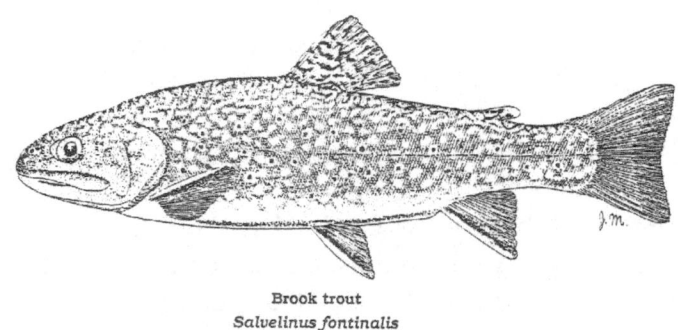

Brook trout
Salvelinus fontinalis

The Brook trout has white leading edges on its paired anal and caudal fins. It has small reddish spots laterally. It grows to about twenty inches.

Sunfish

The "sunfish" are a squat body group of fish that resemble each other so much it makes identification difficult for some people. The fish have jagged back rays that can prick your fingers. They have bottom rear anal fins that resemble boat keels. They all have teeth. The smallmouth, largemouth and rock bass are simply large sunfish. They just outgrow their cousins.

     The smaller sunfish grow from five to seven inches in length. These include Green Sunfish, Pumpkinseed, Warmouth, Bluegill and Longear. The head of the Green sunfish takes up about thirty percent of its body length.

The Pumpkinseed has bright red and orange spots and its cheeks have blue and orange strips. The Warmouth has whitish margins on its gills and teeth on its tongue. The Bluegill has vertical bars on its sides and a dark blotch on its fin just back of the eye. The Long-ear has an extended gill area with a dark spot above its eye level.
    Crappies are a large sunfish in their evolutionary design. They grow to about twelve inches in length and are mostly recognized by their size and shape. They have both beautiful body design and color. The White Crappie has almost indistinct vertical lines and the Black Crappie has dark and light coloring.

If there was a beauty contest for fish, the Black crappie would certainly be one of the finalists. Its color can best be described as dramatic. Below: a twelve inch black crappie.

A Black crappie is a pleasure to view.

Rock bass. Notice the prominent fins. Compare its fins to those of the largemouth bass below.

Longear sunfish have an elongated gill and a dark spot above it. A mature longear might be seven inches long.

Green sunfish may be 6 inches long when fully mature.

Bluegill is every kids favorite fish. They are found in schools and throwing bread crumbs on the water will bring them to the surface. Some grow to ten inches and are good eating if properly deboned. The black spot behind the eye identifies them.

Bluegill are a prolific fish and are often stocked in fish ponds to feed larger predator fish, such as bass. They are fun to catch and the larger bluegill are fun to eat. The states surrounding the Great Lakes do not have limits on bluegill size or numbers that can be harvested.

Research: Fish have sleep-like periods where they have lowered response to stimuli, slowed physical activity, and reduced metabolism but they do not share the same changes in brain waves as humans do when they sleep.

CATFISH

Everyone recognizes a catfish with barbels on its upper lip and lower jaw along with the whiskers on their chins. They do not have scales. The first spine on its dorsal fin can draw blood if you hold it carelessly.
 Catfish go to deep water during bright sunny days and come into the shallow shore at night. So night fishing for catfish would be the most productive.
 Catfish have a tendency to sink rather than float since they have small gas bladders and heavy bony heads. Their flattened head assists them in stirring up mud. Catfish feed by gulping rather than biting or cutting prey.
 Catfish are tolerant of industrial waste and pollution. However, in some industrial areas they have been found with ulcers on their head and lips as well as fin deformation. Eating catfish from Lake Erie can be hazardous to your health. However, catfish in the other lakes are not as contaminated.

Restaurant catfish are almost all raised in ponds. This is a very exacting business since the flavor of the catfish flesh can be affected by its age as well as its type of food consumption. Catfish farms usually employ a "taster" to determine when the fish should be harvested.

The large catfish in the lakes are Black bullhead, Yellow bullhead, Brown bullhead and Channel cat. A somewhat smaller species is the Stonecat. The huge growing Blue Catfish of the Mississippi River is not in the lakes at this writing.
 Most people who fish do not know about the small catfish called "madtom." These only grow to short of four inches in size. They are Mountain madtom, Tadpole madtom, Brindled madtom and Northern madtom.

Mature Black Bullhead are usually around 12 inches. Note the whiskers.

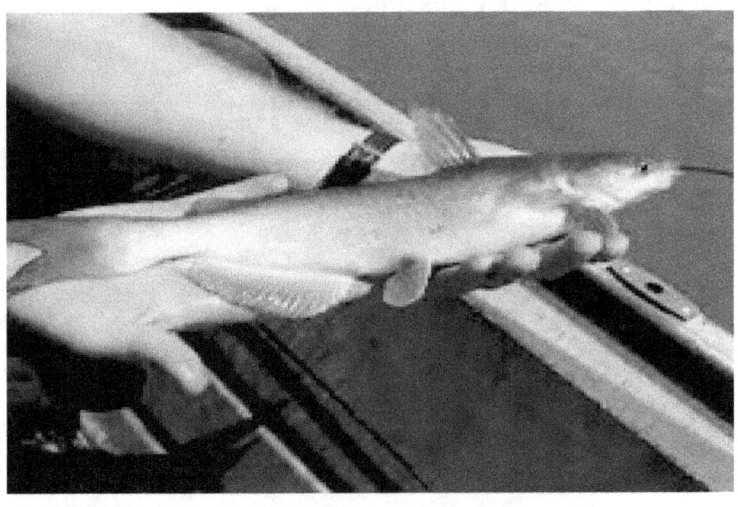

A nice fourteen inch Channel catfish

Some Channel Catfish research

Channel catfish lay their eggs in small cavities dug out of the muddy bottom. Their habits have been extensively studied and are almost science fiction in scope. Channel catfish possess very keen senses of smell and taste. At the pits of their nostrils (nares) are very sensitive odor sensing organs with a very high concentration of smelling receptors. In channel catfish, these organs are sensitive enough to detect several amino acids at about one part per 100 million in water. In addition, the channel catfish has taste buds distributed over the surface of its entire body. These buds are especially concentrated on the fish's four pair of barbels surrounding the mouth.
      This combination of exceptional senses of taste and smell allows the channel catfish to find food in dark, stained, or muddy water with relative ease. Channel catfish also possess an apparatus which amplifies sound waves.

Catfish have enhanced capabilities of taste perception, hence called the "swimming tongue", due to the presence of taste buds all over the external body surface. The catfish has a facial taste system that is extremely responsive to acids. Maybe we don't want to know that much about channel cats and it might interfere with the enjoyment of our catfish sandwich.

A Madtom catfish at maturity.

Suckers and Redhorse

Quillback carpsucker can grow to 30 inches in length. It has a silvery body and a blunt snout.

The sucker group of fish are identified by their thick lower lips and their sucking mouth shape which seems to be on the bottom of their noses. Their heads are without scales and they have an abdominal fin.

Professor Thomas found nine species of sucker-type fish in his survey. These include the redhorse group of sucker which were the Shorthead redhorse, Golden redhorse, Black redhorse and Silver redhorse. All four were found to be present thirty years later in the Toledo University report.

The suckers found were the Spotted, Northern Hog, White, and Longnose. The Quillback is a sucker fish with a different body build than the others. While most suckers have slender looking bodies viewed from the side, the Quillback appears to have a wider body. However, it too, has the sucker mouth.

Quillback carpsuckers have a small sub-terminal (ending below the tip of snout) mouth. The back is moderately arched and the lateral line is nearly straight. They are silvery on the sides with a white belly and the lower fins can be orange or yellow. The tail and dorsal fins are gray to silver in color. The first several rays of the dorsal fin form a long quill. They have no fleshy knob on the front edge of the lower lip which is the characteristic of the other carpsuckers which are mainly found in rivers. Adults are usually 15-20 inches, but can reach 26 inches, and usually weigh 1 to 4 pounds, though they can reach 10 pounds.

The Toledo report also listed the Northern Hogsucker, Quillback carpsucker, Spotted sucker, and White sucker.
Suckers are bottom feeders and are easily mistaken for carp when viewed from above. They are a favorite target of spear fishing. and sometime can be caught with the bare hand along undercut banks in the evening when they rest there. I speak from experience.

A White sucker adult can be twenty inches long. Its head is convex between the eyes which is concave in hog suckers..

The Spotted Sucker is a fish at risk.

Shorthead redhorse adult may grow to 18 inches.

Black redhorse

Longnose sucker

3 Little Fish

The bulk of fish species in the Great Lakes are small in size and are often overlooked by people interested in fish. They are at the bottom of the food chain and are often netted for the purpose of being bait for the larger sport and food species.

The little fish may grow up to five inches in length but most of them are around three inches at maturity. They make interesting fish tank residents. The more abundant of these small creatures are classified as shiners and there are over a dozen types of them in the lakes. There are also different species of dace, sculpin, chub, and flathead minnows. Refer to the Thomas list at the back of this book. index. Other small fish in the lake include darters, stickleback, silverside, killifish, madtom, stone roller and mud minnow.

It is difficult to see the real features of these small fish species since a good detailed illustration would probably be bigger than the fish itself and give the illusion of larger size.

However, get a square bottom net, sink it, put some bread crumbs over it, wait for the little ones to get over it, then pull and when you check the net you will be surprised and delighted.

Chubs

Chubs are probably the most easily identified members of the little fish club because they are large little fish. That's an oxymoron. The larger chubs are good eating if you can get around the bones. Here is what the Encylopedia Britannica says about them.

Chub, any of several freshwater fishes of the carp family, Cyprinidae, common in Europe and North America. Chubs are good bait fish, and large specimens are caught for sport or food.

In North America the name chub is applied to many cyprinids, among them the abundant, widely distributed creek and hornyhead chubs (SEMOTILUS ATROMACULATUS and NOCOMIS, sometimes HYBOPSIS, BIGUTTATA). The creek chub is found in quiet streams in eastern and central North America. Bluish above and silvery below, with a dark spot at the base of the dorsal fin, it grows to about 30 cm (1 foot). It is also called the horned dace, for the hornlike projections that develop on the head of the male during the breeding season. The hornyhead chub is blue-backed with greenish sides and a light belly. It lives in clear streams and is about 15–24 cm (6–9 inches) long. Some chubs will take a fisherman's artificial fly.

Many other unrelated fishes are also called chubs. Certain deepwater lake fishes of the genus LEUCICHTHYS, in the family Coregonidae, are called chub; these chubs are found in the Great Lakes region and are often smoked and sold for food. The rudderfish is also sometimes called a chub, as is SCOMBER JAPONICUS, the chub mackerel found in both the Pacific and Atlantic oceans.

Horneyhead chub

Bluntnose minnow

Dace

Here's what the Encyclopedia Britannica says about Dace.

In North America, the name dace is applied to various small cyprinids. The redbelly daces (PHOXINUS) are well-known, with a southern (P. ERYTHROGASTER) and northern (P. EOS) species. The southern redbelly dace, found in clear creeks from Alabama to Pennsylvania and the Great Lakes region, is an attractive fish sometimes kept in home aquariums. It is 5–7.5 cm (2–3 inches) long and is marked with two longitudinal black stripes. The daces are known for the rosy and bright red coloring assumed by the male during the breeding season. Other North American daces include: the redside and rosyside daces (CLINOSTOMUS), which are black-banded fishes about 12 cm long found in the eastern and central United States. The creek chub is often known also as the horned dace, because of the hornlike projections that develop on the head of the male during breeding season.

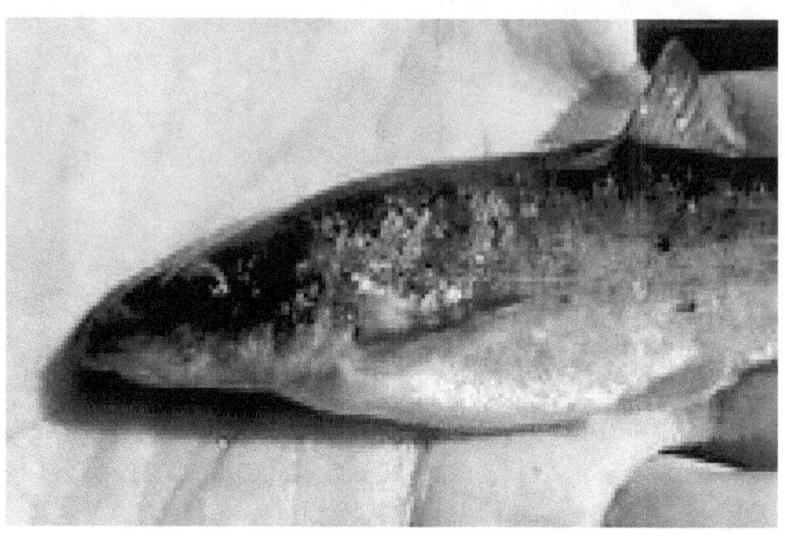

The Longnose dace has a snout projecting beyond its horizontal mouth as well as mottled back and side. It grows to four and a half inches.

A seven and a half inch dace.

Research note: Fish can form schools, containing thousands of fish. They use their eyes and lateral line to hold their places in the school. The lateral line is a row of pores running along the fish's sides from head to tail. Special hairs in the pores sense changes in water pressure from the movements of other fish and predators. The fish on the outside are guided by those in the middle. However, when a predator fish is about there is a mad scramble of the outside fish trying to get into the middle of the school.

My old colleague Raoul Vajk was a Hungarian prisoner of war in Russia during World War I. He taught oceanography and told his classes they, the prisoners, were driven along with whips and everyone on the outside was trying to work their way into the inside.

Shiners

Shiners are a small fish in the genus *Notropis*, except for the Golden Shiner which is in a genus of its own, *Notemigonus*. These make nice aquarium fish but their main use to humans is that of bait fish. The Golden shiner may get as big as seven inches but the others are usually less than four inches in length.

An Emerald shiner has a silver body which reflects a greenish tint. It grows to just over three inches. Very abundant in in the lakes but harvested in large numbers for the sport fishing industry.

The Golden shiner is the most popular of the shiner group of fish. It is golden in its adult stage which is close to seven inches. It is often imitated in the construction of fishing lures.

Minnows

Smaller fish in the subfamily *Leuciscinae* are considered by many to be "true" minnows. However, this is subject to debate since names and classifications are often arbitrary. Below are some discussions of what are considered to be true minnows.

Bluntnose minnow (*Pimephales notatus*): The bluntnose minnow is a primary bait fish for Northern America (more specifically Ohio) and has a very high tolerance for variable water qualities, which helps its distribution throughout many regions. The snout of the bluntnose minnow overhangs the mouth, giving it the bluntnose. There is a dark lateral line which stretches from the opercle to the base of the tail, where a large black spot is located. The average size of the adult is approximately five centimeters or about two inches.

Common shiner (*Notropis cornutus*): These fish are one of the most common type of bait fish and are almost exclusively stream dwellers. The common shiner can be identified by the nine rays on its anal fin and terminal mouth. This minnow is typically bluish silver on the sides and greenish blue on the back., save for breeding season in which case the male gains a rose colored tail and anal fin. The shiner grows to about two inches the first year and maxes out at about four inches aa year later.

Emerald shiner (*Notropis atherinoides atherinoides*): Common shiners are abundant in the Great Lakes. The name of the emerald shiner comes from the greenish emerald band that expands from the back of the gill cover to the tail. This type of minnow has a short, rounded snout, the only difference between the common emerald shiner and the silver shiner is that the silver shiner has a longer snout and a larger eye. These fish grow to an average length of about three inches This is one of the most common bait fish used in the lakes and many fisherman claim it to the best bait for luring predator fish.

Using minnows as bait is no doubt one of the most effective methods for fishing. Bait minnows can usually be found in any bait shop, especially ones near a body of water, but many anglers prefer to capture their own. When we were fishing for walleye or bass we would set out our wire cages in the evening and pick them up early in the morning before we hit the water. Usually we would get enough minnows to last us all day.

A wire trap is made of two half-cylinders with concave cones which end in small openings at either end. These two halves are connected together after some sort of bait to entice the minnows has been placed inside. The wire trap is connected to a length of line which is fastened to something on the bank or a float. Minnows are attracted inside the trap by the bait, swimming into the small opening at either end, after they have found their way in, it is very difficult for them to escape.

Meshnet fishing is a method in which a fisherman finds a school of minnows and using a fine meshnet scoops through the school, bringing the net back out of the water in one motion. This method is primarily used on the shore near the bank of a stream or a shallow area of a lake or pond. I watched a bait shop owner standing in the lake using his meshnet and he had harvested several hundred minnows in about a half hour. He assured me it was legal to do this. I wasn't certain about that and didn't bother to investigate the matter.

To entertain our children we purchased a square dip net with a long handle. We would set the net in a shallow water area where we could observe minnows and we would distribute bread crumbs over it. The minnows and more often bluegill would run in for the bread. A quick pull on the net usually resulted in a lot of small fish to examine.

A seine is a large net, sometimes being 30 or more feet in diameter. It has small weights attached to the bottom. These small weights help to keep the net vertical in the water while two people hold either end of the net and drag it through a spot where minnows are suspected to be. This method is good for catching a variety of bait fish and is one of the oldest methods for catching fish.

One of the most ambitious methods to catch fish I have ever seen was demonstrated on the Canada side of Lake Huron where we had a campsite. The fellow had a boat and a pickup truck and a very big net. He put a stake in the shore and tied one end of the net to it. He had the other end in his small boat with a motor and he made a wide arc in the water and came back to shore where he attached it to his pickup truck and drove away from the lake. Once he got the entire net on shore we went through it and found some debris and weeds, but there was also a lot of fish about four to six inches in length and two very nice largemouth bass.

The Bluntnose minnow is a popular bait fish and as its name implies has a blunt nose which overhangs a small mouth. Its length gets to about four inches when mature.

The Fathead minnow has scales crowded toward the head. Adults have a horizontal dark area across the dorsal fin. This fish is really the definition of minnow since it seldom reaches three inches in length.

4. Lampreys

Lampreys have long eel-shaped bodies and **long dorsal fins** and are without scales. They are also without jaws and have circular, funnel-shaped mouths.

Professor Paul Thomas identified six different species in Lake Erie. These consisted of three parasitic species which include the Ohio Lamprey, Sea Lamprey and Silver Lamprey. When fully grown the smallest of these is the Ohio Lamprey at twelve inches and the largest is the Sea Lamprey at two feet. These often attach themselves to the sides of fish. When the Sea Lamprey made it through the Welland Canal into Lake Erie it decimated the Lake Trout population. Once control methods were enacted, the Lake Trout began making a slow comeback.

The three brook lampreys are Northern Brook, Mountain Brook and American Brook. As their names indicate they can more easily be found in small tributary streams rather than in the lake. In late spring they can be found in clusters on a stream bed sucking up nutrients among rocks. They are around seven inches long and easily handled without fear of being bitten.

The lampreys mentioned above are found in all the Great Lakes. Of the large parasitic lampreys the Ohio Lamprey and the Silver Lamprey are similar in appearance and probably only scientists can tell them apart. Their dorsal fins are continuous. The Sea Lamprey has a separation in its dorsal fin.

A similar situation is found in the smaller brook lampreys where the American brook lamprey has a split dorsal fin while the Northern brook and Mountain brook lampreys have a continuous dorsal fin.

The Sea Lamprey may grow to 25 inches and is identified by its split dorsal fin.

Two lampreys on a host. Can you identify the host?

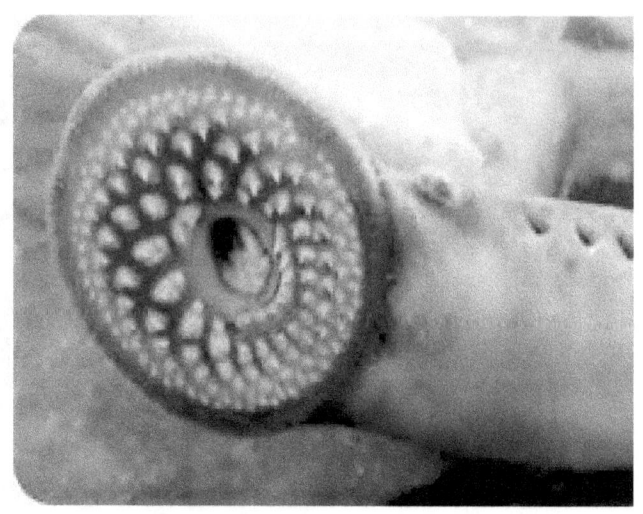

Sucker mouth clamps of a lamprey

The Mountain brook lamprey and the Northern brook lamprey have the continuous dorsal fin and grow to about seven inches. They are easily handled. The American brook lamprey has a split dorsal fin and it grows to about seven inches when mature.

Research: Fish use a variety of low-pitched sounds to convey messages to each other. They moan, grunt, croak, boom, hiss, whistle, creak, shriek, and wail. They rattle their bones and gnash their teeth. They do not have vocal cords and use parts of their bodies to make noises, such as vibrating muscles against their swim bladders. They can also make a sucking sound.

Research: Most fish can see in color and use colors to camouflage themselves or defend themselves and their territory. Most fish have excellent eyesight in their habitat. Some fish can see polarized and ultraviolet light. Goldfish can see infrared radiation. If you keep a goldfish in a dark room, it will lose its color. If you poke your face near the water a fish can see you before you see it. My friend Tom Savko used to crawl on his belly to a fishing sight because he claimed the fish could see him coming. He was a very successful angler.

5. Odd Fish

The Lake Sturgeon has been the source of many lake monster stories. It is an ancient fish species and was once abundant in the lake and many are still caught each year. It can grow to five feet.

The lake sturgeon, also called rock sturgeon, is a North American temperate freshwater fish, one of about 25 species of sturgeon worldwide. Like other sturgeons, this species is an evolutionary ancient bottom feeder with a partly cartilaginous skeleton, an overall streamlined shape and skin bearing rows of bony plates on its sides and back, resembling an arm.

Lake Sturgeon, once in a lifetime.   ODNR photo
It is illegal to have a Lake Sturgeon in your possession so this one was released just after the photo was taken..

The Longnose gar is a most exciting fish to see as well as catch. It will grow to three feet in length.

Gars are called "living fossils" because nearly all their relatives are extinct. They were abundant in Europe during the Tertiary or the first period of the Cenozoic but, before the close of that period, which embraced approximately 58 million years measured by radio-activity, they became extinct in Europe, and the family is now exclusively North American in its distribution.

The gar are also known as billfish and gar pike. They have very sharp teeth. The body length attained may be five feet, but the average is much less. Three-foot specimens are not uncommon. The jaws are elongated into a beak which is twice the length of the head and provided with several rows of teeth which are exceptionally strong, sharp and conical.

Although extremely variable, the color is more commonly greenish above, silvery on the sides and whitish below. The body and fins have large black spots or blotches, and young individuals have a blackish lateral band. The skeleton of the fish is partly cartilage and partly bone.

The spotted gar (L. producrus), which differs chiefly from the Longnose gar in the length of the jaw, occurs in the lakes and is essentially a southern species. Gars may be observed floating like sticks near the surface of the water on warm days or nights. This is a useful form of mimicry by means of which they may drift towards their prey. They are sluggish in their habits except when feeding, when they move swiftly to capture their prey.

The gar possess gills but, because of the fact that the air bladder is connected with the pharynx, it may be used as a lung, and they can rise to the surface to expel air from the air bladder and take in a fresh supply. The ability to use atmospheric air in this way enables the fish to live in waters of low oxygen content.

The gar spawns in late spring or in early summer; they appear in large schools in a suitable spawning area, in close formation in order to assure fertilization of the eggs. The spawn is deposited in shallow, weedy- bays, usually on submerged vegetation or aquatic plant roots.

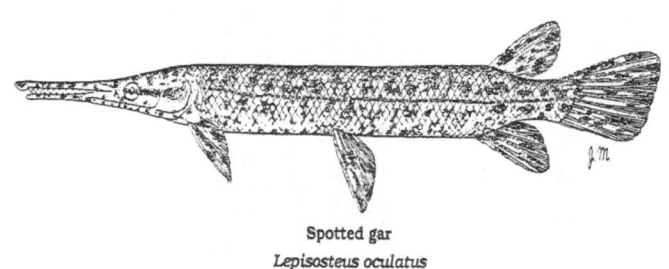

Spotted gar
Lepisosteus oculatus

The Spotted gar has a short and broad snout. The length of the beak is less than 6 times its width. Top of its head has large rounded spots. The adult can reach 36 inches in length.

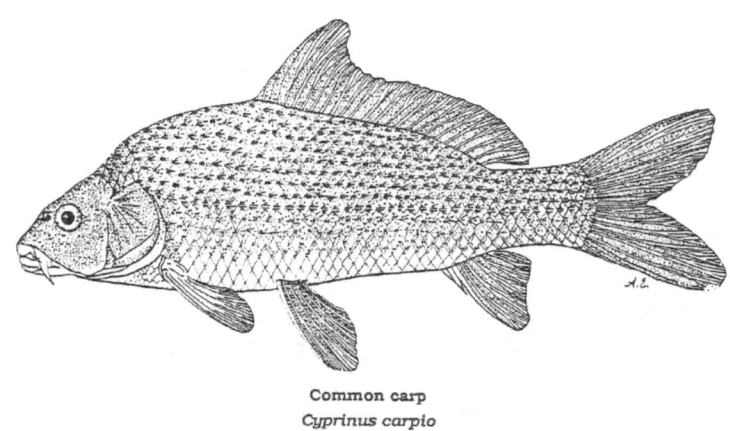

Common carp
Cyprinus carpio

Common carp (Cyprinus carpie)

The Common Carp has been around for a long time. It has two barbels at the angle of its jaws. Its scales have dark spots at the base. The old saying goes to roast a carp on a board, throw away the carp and eat the board. It's funny but not true as carp has some of the best tasting flesh in the fish world. Baking carp frees the flesh from the bones. The biggest carp are often thirty inches in length. The first rays in its dorsal and anal fins are hard and spinous and can cause some problems when not handled correctly.

Common Carp

The Freshwater Drum can reach thirty inches at maturity. It has a blunt snout and what is described as a subterminal mouth. *Note the dorsal fin.*

The Freshwater Drum

The freshwater drum is a good tasting food fish. Though it is a fairly distinctive fish, its deep body, humped back, blunt snout and subterminal mouth have led some to confuse it with the carp and the buffalos. It can be easily distinguished by its two dorsal fins (only one in the carp and buffalos) and its rounded, rather than forked tail. Also, the first dorsal fin of the freshwater drum is composed of 8-9 spines, whereas the carp has only one spine at the beginning of its single soft rayed dorsal fin and the buffalos have no spines at all.

The Bowfin is an ancient fish. Its head is covered with bony plates. It can reach 25 inches in length. Their long dorsal fin and tail structure make them appear ancient, which they are anatomically..

Bowfin (*Amia calva*) are bony fishes. Their common names include mudfish, mud pike, dogfish, griddle, grinnel, cypress trout and choupique. They are regarded as taxonomic relics. They are the sole surviving species of the order Amiiformes which dates back to the Jurassic Period. Although the modern bowfin is highly evolved, it is often referred to as a primitive fish because they have maintained some of the characteristics of their early ancestors..

    The bowfin can tolerate some brackish waters that many other fish find intolerable. They are stalking, ambush predators known to move into the shallows at night to prey on fish and aquatic invertebrates such as crawfish, mollusks, and aquatic insects.

Bowfins, like gars. have the capacity to breathe in both water and air. Their gills exchange gases in water. They also have a gas bladder that allows them to maintain buoyancy and also permits them to breathe air by a connection of a foregut to the gas bladder. They can break the surface to gulp air, which allows them to survive conditions that would be lethal to most other fish species.

Paddlefish – (Polyodon spathula) date back to 300 million years in the fossil record. They were native to the Mississippi River system and have found their way into the Great Lakes. At this time they aare raised in holding tanks and stocked in the lakes.

ODNR photo. Estimate its length and the length of the paddle nose.

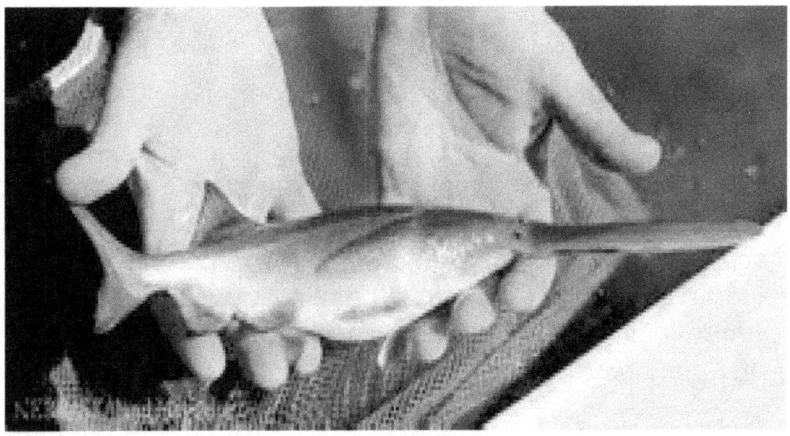

There were no sketches of the paddlefish in the publication of Professor Thomas since they were not abundant in the lake at that time, if they were in the lake at all. It is only through the efforts of various state agencies that paddlefish are found in the great lakes.

American paddlefish are native to the Mississippi River. They had been recorded in Lake Huron in the early nineteenth century. There were no paddlefish recorded in that lake in the early twentieth century. The paddlefish was put under the category of vulnerable endangered species in the 1970s and restricted in international trade.

Paddlefish are long-lived, and sexually late maturing. Females do not begin spawning until they are seven to ten years old, some even as late as sixteen to eighteen years old. Paddlefish spawn in late spring provided the proper combination of events occur, including water flow, temperature, photoperiod, and availability of gravel substrates suitable for spawning. If all the conditions are not met, paddlefish will not spawn. Research suggests females do not spawn every year, rather they spawn every second or third year while males spawn more frequently, typically every year or every other year.

Paddlefish migrate upstream to spawn, and prefer silt-free gravel bars that would otherwise be exposed to air, or covered by very shallow water were it not for the rises in the streams from snow melt and annual spring rains that cause flooding.

We can safely say that all the paddlefish in the lakes are the result of stocking by the bordering state fish commissions and many private enterprises.

The advancements in biotechnology in paddlefish propagation and rearing of captive stock indicates significant improvements in reproduction success, adaptation and survival rates of paddlefish cultured for broodstock development and stock rehabilitation. Such improvements have led to successful practices in reservoir ranching and pond rearing.

Notes of Interest: Fish that have fins with a split tail indicate that they move quickly and may need them to cover great distances. Fish that live among rocks generally have broader lateral fins and larger tail fins.

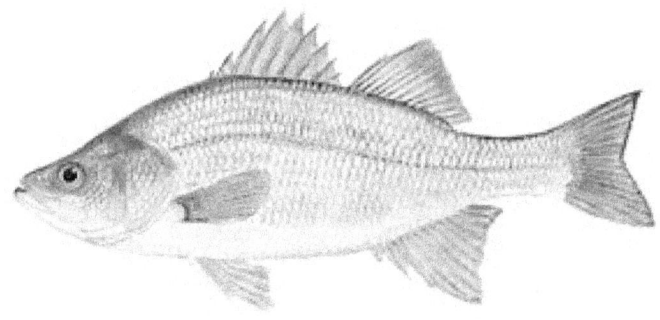

The White perch is classified as a " temperate bass." Its dorsal fins are joined at the base. It has no distinct horizontal stripes on its sides. It grows to twelve inches.

A twelve inch white perch

Point of interest. A ship has a heavy keel in the lower part to keep it from capsizing. Fish have the keel on top. If the paired fins stop functioning to keep the fish balanced, the fish turns over because its heaviest part tends to sink, which happens when it dies.

The term "fish" is used when referring to one species of fish. The term "fishes" is used when referring to more than one species. If you have three trout you can say, "I have three fish." If you have three trout and one carp you would say, "I have four fishes."

A nice smelt seven inches long.

Smelts in the Great Lakes can grow to 14 inches, but few reach that size. They are soft rayed with scaleless heads Their bodies are round. At one time thousands of them would go into the tributary streams to spawn. We would fish them out with dip nets. It was an unbelievable harvest.

Fish research is a branch of zoology known as **ichthyology.** Scientists who specifically study fish are called ichthyologists. There are three distinct groups of fish that ichthyologists study: bony fish, cartilaginous fish, and jawless fish. Ichthyology has a long history that began with simple observations and descriptions of fish over 200 years ago. The first known recorded observations of fish were documented by Pierre Belon in the 1500's. In the early 1700's, Peter Artedi's, who some consider to be the Father of Ichthyology, had his work "Ichthyologia" posthumously published by Karl Linnaeus. At that time, Artedi recognized 230 species of fish. Today, we recognize approximately 31,900 species.

The **round goby** (*Neogobius melanostomus*) is a bottom-dwelling fish, native to central Eurasia which includes the Black Sea and the Caspian Sea. It is believed the Round Goby was transported in a vessel from the Ukraine that had a destination at Lake Huron. Round gobies have established large non-native populations in the Baltic Sea, several major Eurasian rivers, and the North American Great Lakes. It is established in all the Great Lakes.

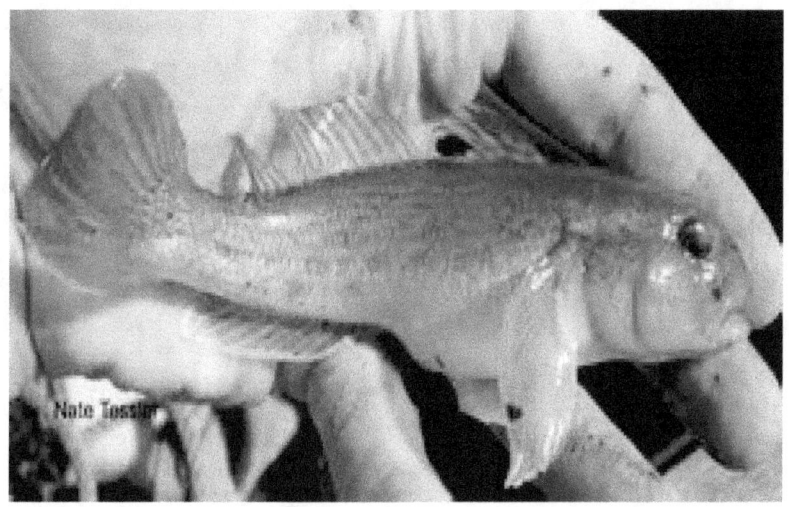

The round goby was first reported from **Pennsylvania** in October 1996, in Lake Erie off Walnut Creek, just west of the city of Erie and later collected in Lake Erie at several locations after that. A study found that several tributaries of Lake Erie had established populations of the round goby. In Erie County, Pennsylvania the round goby was found in Elk Creek, Walnut Creek, Twentymile Creek, and Sixteen Mile Creek. Eventually the round goby was found to be in all of the great lakes.

Also of interest. More than ninety percent of all commercial lip sticks have ground up fish scales in them. Think about that the next time you kiss your sweetheart.

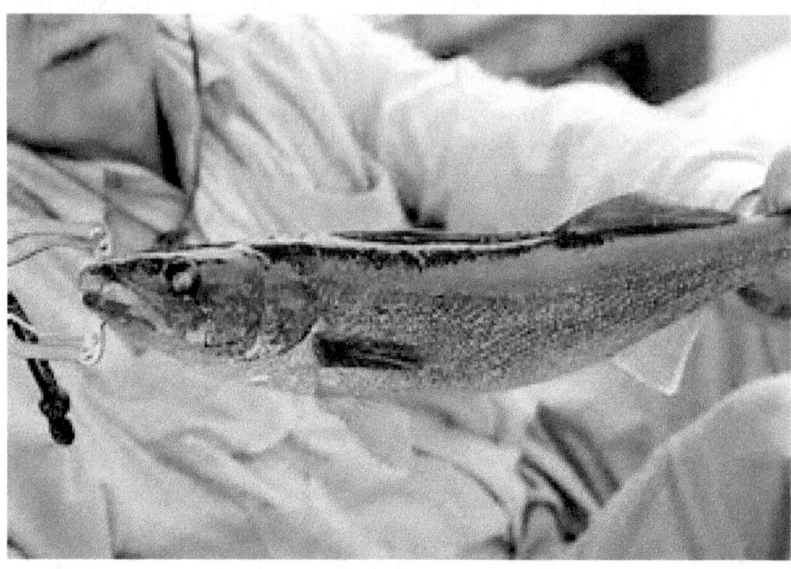

A blue pike. Notice the walleye-like eyes. Those were the good old days.

The **blue pike,** also known as blue walleye, was a foundation fish of the Great Lakes region in the United States and Canada. It lived mostly in the waters of Lake Erie and Lake Ontario. It preferred **cool,** clear waters, living in deep water in summer, and switching to shallow water as they cooled and became less murky in the winter.

The **blue pike** was pursued intensely by commercial and sport fishers, who together landed a billion pounds of the fish between 1885 and 1962. At times, the **blue pike** made up more than 50 percent of the commercial catch in Lake Erie.

The increase in the human population of the region after World War I and the increase in industry and agriculture runoff caused a steady increase in pollution.

All of these actions contributed to the deterioration of the cool, clear habitat needed by the blue pike. During the 1900s, several non-native species of fish made their way to the Great Lakes. These included the sea lamprey, alewife, and rainbow smelt. These contributed to the decline of the blue pike through competition for food and predation.

The blue pike population crashed in 1958, but the **species** lingered on until it became officially extinct in 1970. Other species also became extinct in Lake Erie around that time, notably the three species of cisco. The cisco, however, was thriving in Lakes Huron and Superior.

Evolution at Work

Fish scientists have begun tagging fish in order to determine life spans, growth patterns and perhaps migration routes. In the wild, fish have to survive by cunning and strength. Of course, it pays to have good genes. A prey fish has to win every race.

Wild fish continue to evolve through cross breeding of successful individuals. Unsuccessful fish don't live long enough to breed. This differs from aquarium fish which are manipulated to enhance color, long fins and beauty, all of which are detrimental to the survival of a wild fish.

The young fish, regardless of species, are subject to predation by just about every larger fish in the environment. Whatever the survival rate, those fish that survive are the fish best adapted to the environmental conditions of that place and that time. They are the fish that will continue the species.

The Ohio Survey of 2017 = 52 species identified.

Through the collaboration of several Ohio environmental groups a survey was made of the species of fish found off the Lake Erie coast between Toledo and Cleveland. They will continue monitoring the increase or decrease of the species over the next several decades. The main groups involved in this ongoing study are the Ohio Division of Wildlife, Ohio Environmental Protection Agency, and the University of Toledo.

The following 52 species were identified by their common names -Bigmouth buffalo, Black bullhead, Black crappie, Black redhorse, Bluegill, Bluntnose minnow, Bowfin, Brook silverside. Brown bullhead, Brown trout, Channel catfish, Common carp, Emerald shiner, Flathead catfish, Freshwater drum, Ghost shiner, Gizzard shad, Golden redhorse, Golden shiner, Coldfish, Greater redhorse, Green sunfish, Large mouth bass, Logperch, Longnose gar, Mimic shiner, Northern hogsucker, Northern pike, Orangespotted sunfish, Pumpkinseed sunfish, Quillback carpsucker, Rainbow smelt, Rainbow trout, Rock bass, Round goby, Sand shiner, Shorthead redhorse, Silver chub, Silver redhorse, Smallmouth bass, Smallmouth buffalo, Spotfin shiner, Spottail shiner, Spotted sucker, Trout perch, Walleye, White bass, White crappie, White perch, White sucker, Yellow bullhead, Yellow Perch.

The Fish in Lake Superior – 2014 Minnesota Sea Grant
88 species – includes 34 native species

Common name - Scientific name - Family – N = native species

Alewife – Alosa pseudoharengus – Clupeidae - N
Amerrican brook lamprey – Lampetra lamottei -Petromyzontidae
American eel – Anguilla rostrata – Anguillidae
Atlantic salmon – Salmo salar – Salmonidae
Black bullhead – Ameriurus melas – Ictaluridae - N
Black crappie – Pomoxis nigromaculatus – Centrarchidae - N
Blackchin shiner – Notrois heterodon – Cyprinidae - N
Blacknose dace – Rhinichthys atratulus – Cyprinidae - N
Blacknose shiner – Notropis heterolepis – Cyprinidae - N
Bloater – Coregonus hoyi – Salmonidae – N

Bluegill – Lepomis macrfochirus – Centrachidae - N
Bluntnose minnow – Pimephales notatus – Cyprinidae - N
Brassy minnow – Hybognathus hankinsoni – Cyprinidae - N
Brook silverside – Labidesthes sicculus – Atherinidae
Brook stickleback – Culaea inconstans – Gasterosteridae - N
Brook trout – Salvelinus fontinalis – Salmonidae - N
Brown bullhead – Ictalurus nebulosus – Ictaluridae - N
Brown trout – Salmo trutta – Salmonidae
Burbot – Lota lota – Lotidae - N
Central mudminnow – Umbra limi – Umbridae – N

Channel catfish – Ictalaurus pncatus – Ictaluridae - N
Chinook salmon – Oncorhynchus tshawytscha – Salmonidae
Cisco – Coregonus artedi – Salmonidae - N
Coho salmon – Onchorhynchus kisutch – Salmonidae
Common carp – Cyprinus carpio – Cyprinidae
Common shiner – Luxilus cornutus – Cyprinidae – N
Creek chub – Semotilus atromaculaatus – Cyprinidae – N
Deepwater sculpin – Myoxocephalus thompsoni – Cottidae – N
Emerald shiner – Notropis atherinoides – Cyprinidae – N
Eurasian ruffe – Gymnocephalus cernuus – Percidae

Fathead minnow – Pimephales promelas – Cyprinidae – N
Finescale dace – Phoxinus neogaeus – Cyprinidae – N
Fourspine stickleback – Apeltes quadracus – Gasterosteridae
Freshwater drum – Apolodinotus grunniens – Sciaenidae
Golden shiner – Notemigonus crysoleucas – Cyprinidae – N
Goldfish – Carassius auratus – Cyprinidae
Hornyhead chub – Nocomis biguttatus – Cyrpinidae – N
Iowa darter – Etheostoma exile – Percidae – N
Johnny darter – Etheostoma nigrum – Percidae – N
Kivi – Coregonus kiyi – Salmonidae – N

Lake chub – Couesius plumbeus -Cyprinidae – N
Lake sturgeon – Acipenser fulvescens – Acipenseridae – N
Lake trout – Salvelinus namaycush – Salmonidae – N
Lake whitefish – Coregonus clupeaformis – Salmonidae – N
Largemouth bass – Micropterus salmonoides – Centrachidae – N
Log perch – Percina caprodes – Percidae – N
Longnose dace – Rhinichthys cataractae – Cyprinidae – N
Longnose sucker – Catostomus catostomus – Catastomidae – N
Mimic shiner – Notropis volucellus – Cyprinidae – N
Mottled sculpin – Cottus bairdii – Cottidae – N

Muskellunge – Esox masquinongy – Esocidae – N
Ninespine sticleback – Pungitius pungitius – Gasterosteridae – N
Northern brook lamprey – Ichthyomyzon fossor – Petromyzontidae –
Northern pike – Esox lucius – Esocidae – N
Northern redbelly dace – Phoxinus cos – Cyprinidae – N
Pearl dace – Margariscus margarita – Cyprinidae – N
Pink salmon -Oncorhynchus gorbuscha – Salmonidae
Pugnose shiner – Notropis anogenus – Cyprinidae – N
Pumpkinseed – Lepomis gibbosus – Centrarchidae – N
Pygmy whitefish – Prosopium coulteri – Salmonidae – N

Rainbow smelt – Osmerus mordax, Osmeridae
Rainbow trout – Oncorhynchus mykiss – Salmonidae
Rock bass – Ambloplites rupestris – Centrarchidae – N
Rosyface shiner – Notropis rubellus – Cyprinidae – N
Round goby – Neogobius melanostromus – Gobiidae

Round whitefish- Prosopium cylindraceeum – Salmonidae- N
Sand shiner – Notrois stramineus – Cyprinidae – N
Sauger – Sander canadense – Percidae- N
Sea lamprey – Petromyzon marinus – Petromyzontidae
Shorthead redhorse – Moxostoma macrolepidotum – Catastomidae – N

Shortjaw cisco – Coregonus zenithicus – Salmonidae – N
Silver lamprey – Ichthyomyzon unicuspis – Petromyzontidae – N
Silver redhorse – Moxostoma anisurum – Catastomidae – N
Slimy sculpin – Cottus cognatus – Cottidae – N
Smallmouth bass – Micropterus dolomicui – Centrarchidae – N
Splake – hyrid, lake trout + brook trout – Salmonidae
Spoonhead sculpin – Cottus ricei – Cottidae – N
Spottail shiner – Notropis hudsonius – Cyprinidae – N
Stonechat – Noturus flavus – Ictaluridae – N
Tadpole madtom – Noturus gyrinus – Ictaluridae – N

Threespine stickleback – Gasterosteus aculeatus – Gasterosteridae
Trout perch – Percopsis omiscomaycus – Percopsidae – N
Tubenose goby – Proterorhinus marmoratus – Gobiidae
Walleye – Sander vitreus – Percidae – N
White perch – Morone americana – Moronidae
White sucker – Catostomus commersoni – Catstomidae – N
Yellow bullhead – Aeiurus natalis – Ictaluridae – N
Yellow perch – Perca flavescens – Percidae - N

Professor Paul Thomas 1992 list of species in Lake Erie with their scientific names.

Alewife -Alosa pseudoharengus

Bass, White – Norono chrysops

Bass, Rock – Ambloplites rupestris

Bass, Largemouth – Microptorus salmoides

Bass, Smallmouth – Micropterus dolomieui

Bluegill – see Sunfish

Burbot – Lota iota

Bowfin – Amia calva

Catfish, Black Bullhead - Italurus melas

Catfish, Brindled mudtom – Noturus miurus

Catfish, Brown Bullhead – Ictalurus nobulosus

Catfish, Channel Cat – Ictalurus punctatus

Catfish, Mountain Madtom – Noturus eleutherus

Catfish, Northern mudtom – Noturus stigmosus

Catfish, Stonecat – Noturus flavus

Catfish, Tadpole mudtom – Noturus gyrinus

Catfish, Yellow Bullhead – Ictalurus natalis

Carp, Common – Cyprinus carpie

Chub, Bigeye –Hybopsis amblops

Chub, Creek – Somotilus atromaculatus

Chub, Hornyhead – Mocomis biquttatus

Chub, River – Macomis micropogon

Chub, Silver – Hybopsis storiana

Chub, Streamline – Hybopsis dissimillis

Crappie, Black – Pomoxis nigromaculatus

Crappie, White – Pomoxis annularis

Central Mudminnow – Umbra limi

Dace, Blacknose – Rhinichthys atratulus

Dace, Longnose – Rhinichthys cataractae

Dace,Pearl – Somotilus margarita

Dace, Redside – Clinostomus elongatus

Dace, Southern redbelly – Phoxinus crythrogaster

Darter, Banded – Etheostoma zonnie

Darter, Blackside – Percina maculata

Darter, Channel – Percina copelandi

Darter, Eastern sand – Ammocrypta pellucida

Darter, Fantail – Etheostoma flabellare

Darter, Greenside – Etheostoma bionnioides

Darter, Johnny – Etheostoma nigrum

Darter, Longhead – Percina macrocephala

Darter, Iowa – Etheostoma exile

Darter, Rainbow – Etheostoma caeruieum

Darter, Spotted – Etheostoma maculatum

Darter, Tippecanoe – Etheostoma tippecanoe

Darter, Variegate – Etheostoma variatum

Drum – Apiodinotus grunniens

Gar, Longnose -Lepisosteus osseus

Gar, Spotted – Lepisosteus oculatus

Goldfish -Carrasius auratus

Killifish, Banded – Funduius diaphanus

Herring, Lake – Coregonus artedii

Lamprey, Ohio -Ichthyomyzon bdellium

Lamprey, Northern Brook – Ichthyomyzon tosser

Lamprey, Mountain Brook – Ichthyomyzon greeleyi

Lamprey, Silver – Ichthyomyzon unicuspis

Lamprey, American Brook – Lampetra appendix

Lamprey, Sea – Petromyzon marinus

Minnow, Bluntnose – Pimophales notatus

Minnow, Fathead – Pimophales promeias

Minnow, Silverjaw – Ericymba buccata

Minnow, Tonguetied – Exoglossum laurae

Muskellunge – Esox masquinongy

Muskellunge Tiger – Esox lucius

Perch, Logperch – Percina caprodes

Perch, White – Norono americana

Perch, Yellow – Perca flavescens

Pike, Grass Pickeral – Esox americanus vermiculatus

Pike, Chain Pickeral – Esox niger

Pike, Northern – Esox lucius

Pumpkinseed – see sunfish

Quillback - Carpiodes cyprinus

Redhorse, Black – Moxostoma duquesnei

Redhorse, Golden – Moxostoma erythrurum

Redhorse, Shorthead – Moxostoma macrolopisotum

Redhorse,Silver – Moxostoma anisurum

Salmon, Chinook - Oncorhynchustschawytscha

Salmon, Coho -Oncorhynchus kisutch

Salmon, Sockeye – Oncorthynchus nerka

Sauger – Stixostedien canadense

Shad, Gizzard – Dorosoma cepedianum

Shiner, Blackchin – Notropis heterodon

Shiner, Common – Notropis coronutus

Shiner, Golden – Notomigonus soleucas

Shiner, Emerald -Notropis atherinoides

Shiner, Mimic – Notropis volucellus

Shiner, Redfin = Notropis umbratiis

Shiner, Rosyface – Notropis rubellus

Shiner, Sand – Notropis stramineus

Shiner, Silver – Notropis photogonis

Shiner, Spotfin – Notropis spilopterus

Shiner, Spottail – Notropis hudsonius

Shiner, Striped –Notropis chrysocephalus

Silversides, Brook – Labidesthes sicculus

Sculpin, Mottled – Cottus bairdi

Smelt, Rainbow – Osmerus mordax

Stickleback – Culaea inconstantans

Stoneroller, Central –Compostoma anomalum

Sturgeon, Lake – Acipenser fulvescens

Sucker, Longnose – Catostomus catostomus

Sucker, Northern hog – Hyponielium nigricans

Sucker, Spotted – Minytrema melanops

Sucker, White – Catostomus commersoni

Sunfish, Bluegill – Lopomis macrochirus

Sunfish, Green – Lepomis cyanellus

Sunfish, Longear – Lopomis magalotis

Sunfish, Pumpkinseed – Lepomis gibbosus

Sunfish, Warmouth – Lepomis gulosis

Trout, Broom – Salvelinus fontinelis

Trout, Lake – Salvelinus namaykush

Trout-perch – Percopsis omiscomycus

Trout, Rainbow – Salmo trutta

Trout, Rainbow – Oncorhynchus mykus

Whitefish, Lake – Coregonus clupeaformis

Walleye – Stixostedion vitreum vitreum

Final comment: Recorded and extinct species

The Gizzard shad (Dorosoma cepidimnum) does not appear in the most recent list of fish in Lake Superior. Are we to assume it is not there or perhaps it hasn't been netted by researchers or was it simply overlooked in the listing.
    The Bloater (Corigonus hoyi) was recorded in the 2014 list of fish in Lake Superior, but it has been a long time since one was officially recorded by the fish commissions of Minnesota and Wisconsin. I use the phrase fish commission rather than their proper names for simplification. Its like calling *Salvelinus fontinaulis* a Brook trout.
    When the Cisco was declared extinct in Lake Erie in the 1960s, it didn't guarantee there none were there. It meant that one hadn't been in evidence in the previous ten or more years.

The Blue Pike, (Sander vitreous glaucus) a type of Walleye, was declared extinct in 1970 and none have been recorded in any of the lakes since then. I had caught many of the "blue walleye" back many years ago, but I was less scientific in those days. I believe the species never existed separate from the Walleye. I don't have any evidence for that statement, just that I remember the fish didn't look any different than the Walleye. My fishing companions identified the fish we caught as Blue Pike. The Blue Pike was indigenous to Lakes Erie and Ontario at that time. My opinion is that identifying the blue pike as a separate species due to a color variation is like identifying different humans as separate species based on color.

It would be an interesting study to investigate the actual differences recorded between the Walleye and the Blue Pike.

Probably, a lot of fish species have been incorporated into modern fish lists when actually they have not been in evidence for many years. It would be useful to researchers if a fish commission or university listing indicated the last time a questionable species was "in hand" or in a net.

Of the three lists in the back pages of this work the list presented by the Toledo connection is the shortest and probably the most deserving of attention. It is a listing of fish known to be in the waters between Toledo and Cleveland. It didn't include Muskellunge which makes me uneasy about it. The other lists might be saying, "the Slimy Sculpin has been taken from the lake waters at one time and we are assuming it is still there."

The important research, however, is the determination of the increase or decrease in important desirable recreation and commercial fishing species. I consider boat rentals, lake guiding, and charter boats as commercial fishing, since income to a second party is generated by those activities.

What is important is the study of such things as the ups and downs of the yellow perch population, the decline of the emerald shiner and the health of sport fish as well as a host of other environmental situations. The health of the fish also indicates the health of the lakes and the future of the multi-billion dollar fishing industry that they support.

Asian carp – A threat to the ecosystem

Two species of Asian carp — silver and bighead — were originally imported to control algae in southern catfish farms. Following flooding, they escaped into the Mississippi River in the early 1990s. They now inhabit the Illinois River, which connects to the Great Lakes via the Chicago Sanitary and Ship Canal.

Silver and bighead carp consume vast amounts of food and are extremely prolific. They can weigh up to 100 pounds and grow more than four feet long. Silver carp can jump from the water when agitated and have been known to injure boaters. A moving hundred pound fish can cause a lot of damage.

These species pose a significant threat to the Great Lakes ecosystem due to their large size, voracious eating habits, and rate of reproduction. Natural resource managers fear they would upset the food web, decimate native species, and damage the Great Lakes sport fishery.

It is crucial to prevent Asian carp from entering the Great Lakes. Once established in an ecosystem they are virtually impossible to eradicate. Adult Asian carp have no natural predators in North America and females lay approximately half a million eggs each time they spawn.

Asian carp represent over 97% of the biomass in portions of the Illinois and Mississippi Rivers and are swiftly spreading northward up the Illinois River in the direction of the Great Lakes.

Silver Carp: Hypophthalmichthys molitrix

Bighead Carp: Hypophthalmichthys nobilis

John Tomikel was 50 years old when this photo was taken in 1978.

About the author

John Tomikel graduated from Clarion State Teachers College in Pennsylvania. He became a high school science teacher in the Fairview High School which bordered on Lake Erie. He spent two years in the army during the Korean War, after which he returned to teaching. He continued his education and received a masters degree from Syracuse University and a doctorate from the University of Pittsburgh. He became a professor of earth science at California State College and, later, a professor of environmental science at Edinboro University of Pennsylvania.. He was an outdoor writer for the *Erie Morning News* and had a weekly and Sunday column for fourteen years. He is widely published in the fields of earth and environmental sciences.

Other books on Amazon by Dr. Tomikel include the following:
    Edible Wild Plants and Useful Herbs
    Edible Wild Plants of Eastern U.S. and Canada
    Living with nature at Hawk's Nest
    Earth Processes and Environments
    Environmental Resources Handbook

General Index and page numbers

Illustrations Index page nos.

Alewife 12, 19
Asian carp 146

Banded darter 63
Banded killifish 54
Bigeye chub 34
Black bullhead 49,101
Blackchin shiner 37
Black crappie 58, 96
Blacknose dace 40, 42
Blackside darter 69, 84
Black redhorse 46,107
Bloater 73

Bluegill 60, 99
Blue pike 133
Bluntnose minnow 41,109,115
Bowfin 18, 126
Brindled madtom 51
Brook lamprey 118
Brook silverside 54
Brook stickleback 55
Brook trout 95
Brown bullhead 50
Brown trout 20, 93
Burbot 53

carp 123, 123
Central stoneroller 32
Central mudminnow 24

Chain pickerel 25, 80, 83
Channel catfish 48,101
Channel darter 38
Chinook salmon 21
Cisco 23, 75
Coho salmon 22, 90
Common shiner 38
Creek chub 39

dace 111
drum 71, 124

Eastern sand darter 67
Emerald shiner 37, 111
Fantail darter 65
Fathead minnow 41, 115

gar 121, 125
gizzard shad 13, 19
goby 76, 130, 131

Goldfish 28
Golden redhorse 46
Golden shiner 36, 112
Grass pickerel 25, 79
Greenside darter 66
Green sunfish 61, 98

herring 23, 89
Horneyhead chub 35,108

Iowa darter 65

Johnny darter 64
Killifish 54

Largemouth bass 57, 97
Lake herring 23, 89
Lake trout 92, 94
Lamprey 117, 118
Logperch 68
Longear sunfish 59, 98
Longhead darter 70
Longnose dace 39, 110
Longnose gar 17, 120
Longnose sucker 42, 106

Madtom 102
Mimic shiner 30
Mountain madtom 50
mudminnow 24
Muskellunge 27, 78

Northern brook lamprey 16
Northern madtom 52
Northern pike 26, 79
Northern hog sucker 44

Paddlefish 126. 127
Pearl dace 40
Quillback 44, 103

Rainbow darter 66
Rainbow smelt 23, 131
Rainbow trout 20, 93
Redfin shiner 29
Redside dace 32
River chub 36
Rock bass 61, 97

Rosyface shiner 30
ruffe 74

Sand darter 67
Sand shiner 42
Sauger 69, 84, 86
Sculpin 72
Sheepshead 71
Shorthead redhorse 43. 105
Silver lamprey 16
Silverjaw minnow 33
Silver shiner 31
Silversides 55
Silver redhorse 45

Smallmouth bass 59
Smelt 23, 130
Southern redbelly 29
Splake 73
Spotfin shiner 31

Spotted darter 64
Spotted sucker 45, 104
Spottail shiner 35
Sockeye salmon 21, 91
steelhead 20, 94
stickleback 55, 75
Stonecat 51

Streamline chub 34
Striped shiner 28
Sturgeon 17, 120

Tadpole madtom 48
Temperate bass 56
Tiger muskellunge 26

Tippecanoe 63
Tonguetied minnow 33
Trout-perch 52, 86
Tubenose goby 77

Variegate darter 67
Walleye 70, 85
Warmouth 60
White bass 56
White crappie 58

whitefish 22, 89
White perch 56, 128
White sucker 47, 104
Yellow bullhead 49
Yellow perch 62, 84

www.ingramcontent.com/pod-product-compliance
Lightning Source LLC
Chambersburg PA
CBHW070246230526
45470CB00002B/492